The BEST from AMERICAN CANALS Number VI

INTRODUCTION

It has now been twenty-one years since the American Canal Society was formed, and since it published its first quarterly Bulletin. In August of 1993, we published Bulletin Number 86, providing two more years of excellent feature articles sent in by our ACS members since the last "BEST" was published in October of 1991.

This Sixth "BEST" is an 88-page Journal in exactly the same format as its five predecessors, all of which are still available. Reviewing all six publications, we find that there is very little duplication among them. Together they form a highly detailed encyclopedia of canal and waterway information from around the world.

All these articles, including their hundreds of photographs, maps and drawings have been carefully researched by our members to make them as accurate as possible. Nothing like this treasure-trove of waterways information, historic and current, is available in such concentrated form – anywhere else!

York, Pennsylvania
November 1993

William H. Shank, P.E.
Publisher, AMERICAN CANALS

COVER ILLUSTRATION

The Allegheny Portage Railroad was part of the Main-Line Canal System conveying travelers and freight between Philadelphia and Pittsburgh in the early 1800's. A sectional canal boat train is shown in this Hoffmann drawing, crossing the Edensburg Viaduct of the Portage at Wilmore.

Published by
The American Canal and Transportation Center
809 Rathton Road, York, Pa. 17403

ISBN 0-933788-85-1

TABLE OF CONTENTS

A VISIT TO LOWELL

By Bill Shank

During the week-end of May First and Second, 1993, I was the guest of our esteemed ACS Vice President, Bill Gerber, of North Chelmsford, Massachusetts. In this two-day period Bill Gerber gave me an accelerated education in the Saugus Iron Works, the Charles River outlet locks at Boston, the Baldwin Mansion and the waterways system at Lowell, Massachusetts, including the Merrimack River, the Concord River, the Pawtucket Canal and the Middlesex Canal.

My main objective was to visit with the members of Middlesex Canal Association, at their annual meeting May 2nd, at the Wilmington Arts Center in Wilmington, Massachusetts, as their guest speaker. My subject: "The Amazing Pennsylvania Canals", which I presented as a slide show. I found a number of familiar faces in the audience, included Dave Barber and his wife; Col. Wilbar Hoxie and his wife, not to mention Bill Gerber himself, who was responsible for my being there.

The audience was most generous in their comments about my presentation and bought a quantity of *PICTURE JOURNEYS ALONG THE PENNSYLVANIA MAIN LINE CANAL*. Some signed up on the spot as members of the American Canal Society.

I managed to make a few photos during the week-end, some of which are shown here. I am deeply indebted to both Bill Gerber and the Middlesex Canal Association for their cordial reception. It just proves the old adage: "An expert is anyone 500 miles from home".

Several members of the Middlesex Canal Association pose in front of the "Colonel Baldwin" canal boat replica, which operates in the summer along a restored section of the Middlesex.

The Baldwin Mansion (now a fine restaurant) was the site for our dinner after the annual meeting of the Middlesex Canal Association. The Baldwin House was the home of Loammi Baldwin, Chief Engineer for the MIddlesex Canal.

A boat passes through one of the tide-water locks on the Charles River in downtown Boston.

Bill Gerber next to the "cow-catcher" on the fully-authentic replica of an early street car which carries visitors around the Lowell Heritage State Park.

Restored lock on the Pawtucket Canal near the Sheraton Inn in downtown Lowell.

Bill Shank and Bill Gerber pose for the cameraman in front of the "Colonel Baldwin".

ADAPTING IN A CHANGING WORLD

September 29–October 2, 1993 — APT XXV CAN — Adapting in a Changing world, the Association for Preservation Technology International's 25th anniversary conference, will be held at the Chateau Laurier in Ottowa, Canada. The conference will bring together architects, engineers, conservators, historians, craftspersons, preservationists and other professionals involved in the maintenance and preservation of our built heritage. Training courses on *Historic Roofing* and *Masonry* and a colloquium on *Conservation Management* precede the conference on September 26–29, 1993. Among the unique preservation challenges is the ongoing repair and operation of the historic Rideau Canal system. For additional information, contact Robert Hunter, APT CAN Conference Chair, at (819) 997-6974 or FAX (819) 953-4909.

LOWELL, MASSACHUSETTS: RESTORING A CITY OF CANALS

The restored Suffolk Mill building, originally built in the 1850s. Power canal in foreground runs into it and turns turbines.

by Bruce J. Russell, Contributing Editor

Situated 30 miles northwest of Boston, on the Merrimack River, Lowell was once one of America's foremost industrial cities. Textiles were its mainstay, and during the Civil War, uniforms for the Union army were produced by the hundreds of thousands. Enormous multistory brick buildings dotted the landscape, and the sound of countless looms and knitting machines could be heard from 6:00 a.m. until dusk. Laborers toiled at their looms 12 or more hours a day, and the sight of minor children working alongside their parents was a common one.

Labor was also provided by young women who had been raised on nearby farms and were sent to the mills to earn income for their families. Their lot was not an enviable one. After they had finished their long day of work, the women, some less than 16 years of age, retired to company-owned dormitories and boarding houses where conditions were more akin to reform schools than places of domicile. This was the "Lowell System," and represented the darker side of America's emergence as an industrial power. By the middle of the 19th century, feminist notions had begun to develop among the intellectuals of New England. Opposition to the idea of subjecting young Yankee maidens to such degrading conditions took root, and fewer and fewer applied for work in the mills.

Nevertheless, the need for labor continued, and the firms began to look overseas. Waves of immigrants came to work in Lowell for wages that allowed a bare bones existence. In the 1850s they originated in Wales and Scotland, but a decade later a majority were impoverished Irish peasants. Still later in the 19th century, eastern and southern Europeans streamed into Lowell to labor in the textile mills. Prosperity for the owners lasted until the end of World War I, after which the manufacture of clothing, shoes and hats began to shift to the Carolinas and farther south, primarily to avoid the wage demands of increasingly unionized New England workers.

Throughout the 1920s, closures occurred, and the once-busy mills became deserted. People left Lowell, and parts of the city began to take on the appearance of a ghost town. Many of the grandiose mansions of the "textile barons" were sold and subsequently converted into rooming houses. What had earlier happened to New England whaling centers, as petroleum replaced sperm oil, began to occur throughout the Massachusetts textile region. The Great Depression began in Lowell years prior to its appearance in the rest of the nation. By 1930, over two thirds of the multistory brick mill structures were permanently boarded up, with the sign "NO WORK" posted outside. Economic stagnation continues today, with many poor people inhabiting once-prosperous neighborhoods.

Lowell industries continued to rely on water power for decades after the development of the steam engine. The city's location on the swift-moving Merrimack River allowed the digging of power canals which brought water directly to the sites where mills were constructed. Water wheels and eventually turbines were placed where canals ran through basements of the factory buildings; gears and belts carried the rotary power to the machines on each of the four or five floors above. The Pawtucket Canal, dug in the 1830s and subsequently enlarged, drew water from the Merrimack and fed it to the multitude of power canals through an intricate system of guard locks and sluice gates. In addition, the Pawtucket Canal, unlike its tributaries, was a navigable waterway. Raw materials could be transported into the city and finished products removed aboard barges.

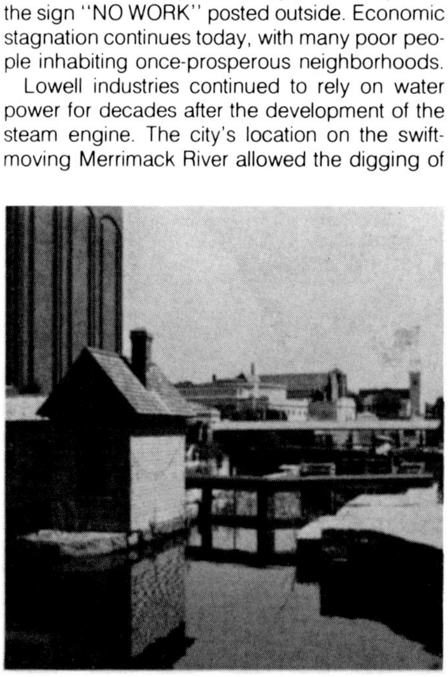

One of the gatehouses, containing mechanism for controlling water, and adjacent lock chamber to permit vessels to descend to lower levels. Restored during the 1980s from old plans and diagrams.

All the water diverted from the Merrimack was returned to it after being used for manufacturing and transportation. The system exhausted no natural resources and caused no pollution. Only when electric power became an ecnomically feasible alternative did the canals and turbines begin to be abandoned. Electricity had the great advantage of flexibility, with each machine being independently powered. In addition, the system of fast-turning shafts moving thick leather belts was inherently dangerous. Many workers were killed or maimed after becoming entangled. It was impossible to stop a machine except by disengaging a belt or shutting down the entire factory. However, even after the introduction of electric motors, many of the mills retained their turbines and some belt-driven looms well into the 20th century. If power outages occurred, turbines could be reactivated and limited production continued.

Lowell's last mill went out of business after World War II, and the city languished through the 1950s, 60s, and 70s. Hispanics replaced earlier European immigrants, and poverty replaced affluence. A scheme to dynamite most of the remaining mills and create vast open spaces was proposed during the 1970s by the Commonwealth of Massachusetts as a means of doing away with an eyesore standing in the way of possible urban renewal. However, a number of people interested in industrial archaeology began to organize a coalition to prevent this from happening. They were aware that Lowell, with its unique canal system, symbolized Yankee ingenuity at its best, and believed that what was left should be retained and restored, if possible. Their argument was simple: a story exists in this old mill town, and future generations need to hear it told through actual contact with all of its elements. What better way to do this than to make Lowell a living museum of mid-19th century New England and American industrial life and technology? To them, Lowell was a city run by water power transmitted through man-made canals—there was none other like it in the United States. Ultimately, they convinced both the Massachusetts and the federal government that it would be reckless and irresponsible to demolish all this simply to create downtown parking spaces and fast-food restaurants.

The canals were weed-grown and filled with garbage, the mills were deteriorating firetraps, and the turbines were rusted hulks, but if enough dollars were spent, they could be restored. In effect, the scheme was to turn a poor city into an urban park, where the story of New England textile manufacturing could be told through interpretive exhibits and rebuilt structures including canals, locks, sluice gates, turbines, and other kinds of machinery. Eventually, the Governor of Massachusetts and enough legislators saw the wisdom of revitalizing Lowell, and all plans for demolition were shelved in favor of an ongoing restoration in the heart of the mill district and the major navigational and power canals connected to it. Initial efforts were aimed at stabilization, to prevent further deterioration; following that, a complete master plan could be formulated. The ultimate goal was a recreation of Lowell's core as it was in the 1860s, when water rather than steam or electricity kept the wheels of its machines turning. In addition, the entire story of textiles would be told as well as that of immigrant and child labor. There would be exhibits on dyeing fabric and stitching it into shirts, pants, and other items. Looms of various types would be described as well as the design of various kinds of water-powered turbines and gears. Belt drives and bobbins were likewise to be explained, so a visitor might gain a complete knowledge of what went on in Lowell over a century ago. Industrial archaeologists from the Smithsonian Institution provided valuable expertise, sometimes furnishing rare photographs and blueprints from their extensive collections. The ultimate goal was historical accuracy rather than a "theme park" type of setting.

By 1980, the Lowell National Historical Park and the Lowell Heritage State Park had been created, the former under federal auspices and latter administered by Massachusetts. Funding from both federal and state levels was therefore assured. Since this time, the National Park Service has been responsible for the actual running of the park, and uniformed rangers conduct a number of informative tours. Initially, there wasn't much to see, since Lowell was a virtual shambles after decades of neglect, but during the past 12 years a great deal has been accomplished. Guard locks, gatehouses, canal walls, and other features of 19th century Lowell are again in working condition. The Suffolk, Boott, and Market multistory mill buildings are fully or partially rehabilitated. Furthermore, at least one power canal and its turbine can be seen, and ultimately it will be set up actually to run several looms. Enormous research, including locating original plans, made these restorations possible. Much more needs to be done, but an excellent start has occurred. Lowell each year attracts many visitors, and is included in the itineraries of

Mechanism for controlling flow of water on one of Lowell's canals.

specialized tours of 19th century industrial America. The guest register at the visitors' center shows names and addresses from all over the world.

A boat ride is offered on the Pawtucket Canal, which takes visitors past old mill buildings, both restored and awaiting refurbishment as funding becomes available. A trip through a working lock is also featured. The park ranger explains how a lock functions as well as what the rôle of a gatehouse was. Diagrams given to the passengers show the layout of Lowell's canals, and how the water was diverted to each mill. In addition, visitors to the "Spindle City," (as Lowell was called,) can take a trip on a replica 1903 streetcar, which also passes historic sites. This line is at present being extended. A visitors' center is situated in the partially restored Market Mill complex, and here one can see films and view displays of life in old-time Lowell. A book store offers a wide range of titles relating to canals, industrial archaeology, and the textile industry. An award-winning slide show, "Lowell: The Industrial Revolution," can also be seen. At the restored Boott Mill, there is now a room filled with 90 operating looms which were powered by water. In the mid-19th century, all of this was considered high tech.

The canal enthusiast will enjoy a trip to Lowell, including a ride on the Pawtucket Canal. Cruises operate in July and August. The two-hour Sunset Cruise is especially recommended.

Restored lock chamber and gates in Lowell.

New Canal Boat Construction Index

The Canal Boat Committee of ACS, Chairman William J. McKelvey, Jr., is pleased to announce the publication of the 1993 Canal Boat Construction Index, Third Edition. Previously published in 1975 and 1979, the Index has been edited by Carroll Gantz, to provide a comprehensive reference for the design and construction of canal boats authentic to their time and place in American History.

The Index contains all known information relating to the design and construction of canal boats, including books, descriptions, photographs, drawings, remains, models, plans, re-constructions, planned re-constructions, and potential sites.

Visitors to Williamsburg, Plimoth Plantation, The U.S.S. Constitution, Ft. McHenry, Kitty Hawk, Greenfield Village, or any other well known restoration of American historical sites or artifacts; would be understandably disappointed to find a "Pilgrim" wearing Reeboks, plywood interiors in sailing vessels, or plastic replicas of barrels in a "forge". Yet we see replicas of canal boats with fiberglass hulls, wood trim in 2" X 4"s, plywood "decking", plastic "canvas", outboard motors, hardware from Builder's Square, and other examples of insensitivity to the authentic materials, construction techniques, finishing, and operations of the Canal era.

When questioned about these anomalies, well-intentioned people often cite the extra cost, time, or difficulty in obtaining authentic traditional materials or craftsmanship. Often they are concerned with the time and cost of maintenance of traditional materials, and sometimes they just feel that the "boat ride" is the main thing, and that the details of the actual materials, construction, and operation are not of interest to the public.

Carroll Gantz, an ACS member who is also a professional Industrial Designer, responds to these issues by reminding planners that any historical reconstructions are not only recreational, but also educational and emotional in nature. The design objective is to re-create an emotionally and educationally meaningful total historical *experience* for the visitor. This is accomplished by the sights, sounds, facts, and feel of mid-nineteenth century technology; created by many small details reinforcing each other. The design of the boat and it's operation should be appropriate for that particular geographic location and history. Each site is unique. There could be nothing more ridiculous than putting an Ohio canalboat design on the C&O canal in Maryland, for example.

A copy of the Index is being mailed to all members of ACS, and additional copies are available from Keith Kroon, ACS Sales, 2240 Ridgeway Ave., Rochester, NY 14626.

GENESEE VALLEY CANAL CHRONOLOGY

Black Creek Aqueduct/Culvert—This beautifully preserved large double arch structure spans a major tributary to the Genesee River about 5 miles south of Rochester. It served Genesee Valley Canal Railroad, later a branch of the Pennsylvania Railroad for nearly 100 years following its abandonment by the Canal in 1878.

Compiled by David L. Kipp

SIGNIFICANT DATES

1823: 1st petitions to improve navigation of the Genesee River received by the NYS Legislature.

1825: Bill to authorize a survey of a canal route between the Erie Canal and the Allegany River introduced in the Legislature but not passed.

Public meeting called in Geneseo to promote a Genesee valley canal.

1827: Gov. DeWitt Clinton recommended investigation of navigation from the Erie Canal to the Allegany River.

1828: A preliminary survey of the potential GVC route was made under the direction of Judge James Geddes.

1830: The Legislature appropriated $750 for a "careful" survey (not done).

Meetings of GVC supporters and promoters held in Dansville, Angelica and Geneseo.

1833: The GVC route to the Ohio River Valley was extolled in the "Annual Register and Universal Gazetteer".

1834: The Legislature authorized a detailed survey of the GVC route to include a side-cut to Dansville.

1835: Detailed survey conducted under the direction of Frederick C. Mills.

1836: Frederick C. Mills (appointed Chief Engineer) submitted a survey report. He recommended a canal on the west side of the Genesee River. It would be about 125 miles long (107 miles on the mainline from Rochester to Olean, 7 for the extension from Olean to Millgrove on the Allegany and 11 for the side-cut to Dansville). It would cost $1.9 million and require about 1060 vertical feet of lockages, 15 aqueducts, 8 dams, 134 culverts, 103 highway bridges and 130 farm bridges. A 1000 foot tunnel would be required at Portageville in the vicinity of the 3 falls in what is now Letchworth State Park.

111 petitions were received by the Legislature urging construction of the GVC.

In May, a law was passed authorizing construction of the GVC.

1837: In June, the first contracts were let (except for water supply structures) and construction was started. Anticipated completion date was 1842.

By November, 28 miles were under contract.

1838: 50 more miles put under contract.

Two tunnel plans were considered for negotiating the Portageville Moraine between Oakland and Portageville (Rt. 436 "ups and downs"). The long tunnel plan was initially preferred because it would have shortened the route by 1 3/4 miles in eliminating a loop through what is now the SE part of Letchworth State Park opposite the Middle Falls/Glen Iris Inn area. It would have involved a 1720 ft. tunnel directly through the moraine. However, test borings and test pits determined that the moraine would not support a tunnel. The short tunnel plan was adopted which was to punch a 1080 ft. passage under the 100 ft. high hill just south of Middle Falls and opposite what is now "Inspiration Point". Elisha Johnson of Rochester is awarded the tunnel section contract.

1839: In January, 1st 2 miles were completed to "Rapids Settlement" across the Genesee River from the present UofR river campus. It was decided that locks are to be 15 feet wide by 90 feet between gate quoins and made from hammer dressed stone. The prism is to be 42 ft. wide at the surface, 26 ft. wide at the bottom and 4 ft. deep. These lock and prism dimensions are the same as the original Erie Canal (Clinton's Ditch).

The "Deep Cut" through the crest of a hill between Nunda and Portageville is started. It will be 1000 yards long, up to 73 feet deep and 285 feet wide at the top. 587,000 cubic yards of dirt eventually removed.

The tunnel project at Portageville is started and 460′ of the excavation are completed. Dimensions are 27 feet high and 20 feet wide.

In June, cheaper locks (composite or wood) are authorized. This was estimated to save $385,000. Not counting the mud lock and 4 guard locks, 112 locks were required: 28 of stone, 73 composite (rough stone with plank lining of the chamber) and 11 wood.

Final contracts were let.

1840: Erie Canal junction (Rochester) to Mr. Morris opened. Sept. 1st, first boat arrived. Daily packet boat service started.

Hornby Lodge constructed by Chief Engineer Elisha Johnson over the tunnel site at Portageville to be his home during the project. 760′ of the tunnel now complete.

Scottsville Feeder Gates—Allen's Creek (Now Oatka Creek) waters were impounded by an arched dam just to the right and sent down the Scottsville Feeder through these gates. This was one of four sites where the Genesee River was tapped for water for the canal. Iron gatepost collars are still in place.

Portageville Aqueduct Site—Having completed its diversion to the east side of the Genesee River to bypass the Letchworth Gorge, the GVC crosses over to the west side again over a 7 span, wooden trunk aqueduct 50 ft. above the river. The abutment on the far side is preserved unchanged from the canal era. The piers in the foreground are remnants of the RR trestle that replaced the aqueduct.

Collectors office established in Scottsville. $6900 collected the first year.

Property value of the canal right-of-way listed at $400,000.

1841: Canal open to Dansville—now a total of 51 miles open.

1842: Tunnel ready for masonry arching.

1842 to 1847: All canal construction work in New York State stopped—credit crisis. Unfinished at this time: Shakers (Sonyea) to Olean—65.5 miles. 95 locks remained to be completed—71 foundations for these had been laid.

1847: Works resumed (slowly). Part of tunnel ceiling found to have collapsed. Bids were solicited for resolving tunnel area dilema: Complete tunnel with masonry arching—$104,000. Open cut around the edge of the hill on the edge of the gorge—$32,000.

1848: Portageville tunnel abandoned for an open cut after spending $250,000 on it. The abandoned tunnel became the home of a large bat colony and a destination for intrepid explorers until the Park recently sealed off the entrance for safety reasons.

1849: Work continues slowly along the canal line due to inadequate funding, strikes and the beginning of a competitive thrust from the Erie Railroad entering western New York.

1851: Canal opened from Mr. Morris to Oramel, just south of Houghton Creek, site of "Jockey Street".

Erie Railroad completed to Olean.

Small Culvert—A well preserved 4 ft. arch culvert a few miles south of Mt. Morris, NY. The Canal Prism is readily visible here for several miles paralleling NY Route 36.

1853: Canal opened to Belfast.

1854: Canal opened to Rockville.

Canal right-of-way value listed at $5.4 million.

Tonnage carried by the canal peaked this year: 159,000 tons. Boat departures and arrivals at Rochester totaled 5350.

1856: First boat arrives in Cuba.

Canal completed to Olean.

1857: The Legislature authorizes the extension from Olean to Mill Grove.

1858: Oil Creek Reservoir (now Cuba Lake) completed. Provided water to the summit level. Dam is 2000 feet long, 60 feet high, 290 feet wide at the base; the largest in the world at the time. It created a lake of 500 acres, 1665 feet above sea level.

1859: 6 of the 6.7 mile extension from Olean to Millgrove Pond on the Allegany River and all but the last two locks are completed.

1861: Extension to the Allegany River at Millgrove completed. Total construction costs about $6 million, 3 times original estimates.

1862: First navigation season with completed Canal.

Summit level closed for 60 days this year and in 1863 due to lack of water.

1868: Oil Creek Reservoir dam raised 6 feet. Ischua Creek dam raised 2.5 feet.

1869: Oil Creek Reservoir dam raised 2 more feet and Ischua Creek another 6.5 feet. This allowed 27 boats per day to pass over the summit for the entire navigation season.

Note: After the GVC was closed, the Oil Creek Reservoir Dam was lowered to its original height due to public nervousness in regard to man-made dams following the Johnstown Flood disaster in 1889.

1874: Opening of navigation delayed until June because of winter ice and spring flood damage, primarily in the Keshequa Creek valley.

NYS authorized to sell or lease canals not financially viable.

1875: Canal Commissioners determined which lateral canals should be leased, abandoned or retained as feeders.

Aggregate cost of GVC construction and improvements to date is $6.7 million.

(Concluded on Page 8)

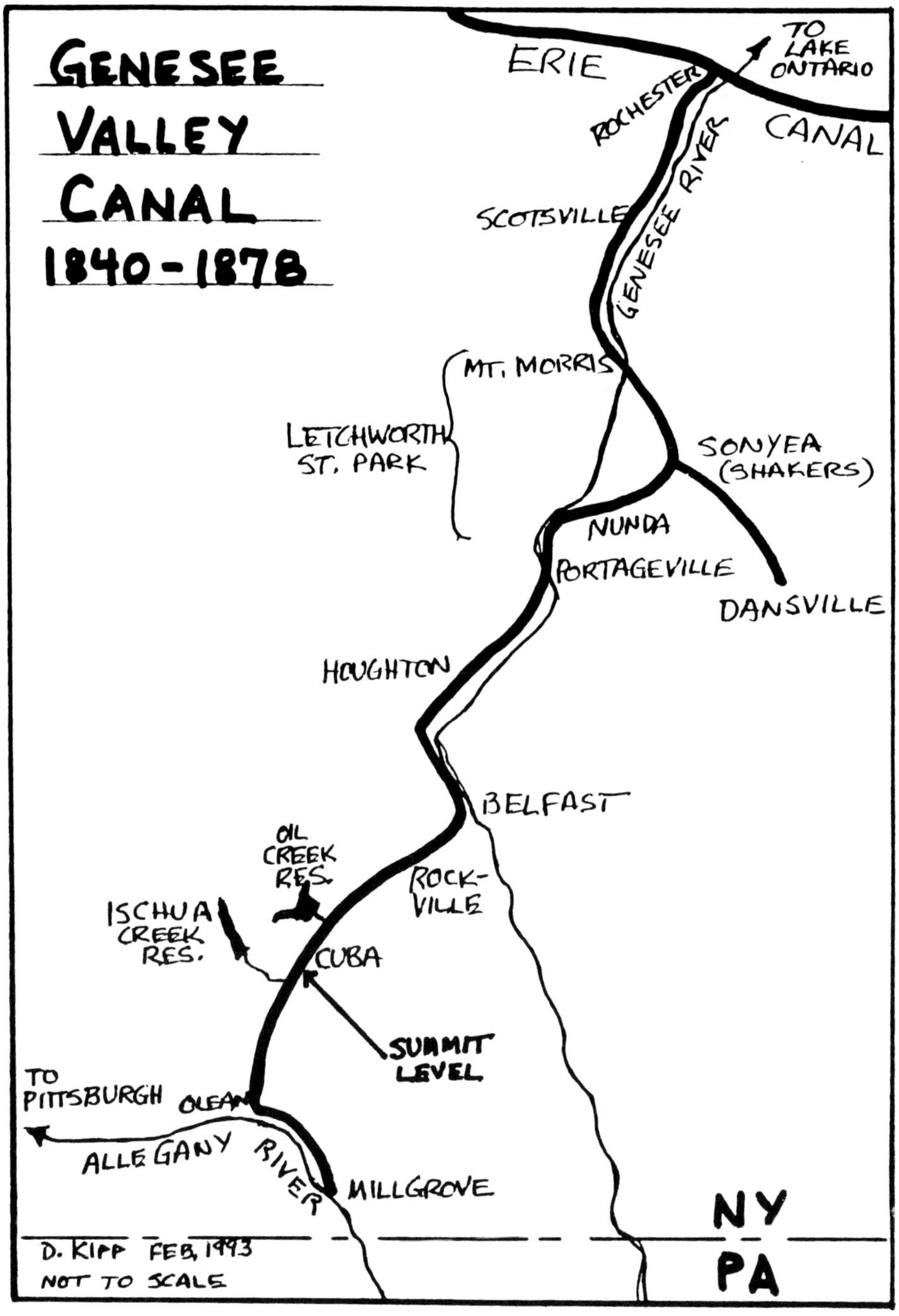

1792 BATTEAU RECREATED

The "Discovery" — replica of an Eighteenth Century Batteau, such as used on the Mohawk River and early waterways west of Albany. Overall length: 30 feet, 4 inches. Maximum width at the gunwales: 7 feet, 3 inches. Carrying capacity: 1.5 tons.

ALBANY, N.Y. — New York State Museum technicians are building the first replica ever constructed of an 18th century Mohawk River batteau as part of a two-year statewide celebration of the birth of the Canal Era in New York. The batteau, named DISCOVERY, will be on exhibit at the Urban Cultural Park Orientation Center at the Schenectady Museum throughout the summer, and will participate in reenactments and living history programs planned at towns along the modern Barge Canal.

"Batteau" is a French term for "boat" which came to signify, 200 years ago, any flat-bottomed, shallow-draft vessel that was pointed at both ends. This vessel was the mainstay of inland shipping, particularly for the military, until the end of the 18th century.

Batteaux (the plural) came in different sizes, known generally as 3-handed, 4-handed or 5-handed according to the crew needed to propel them. There were undoubtedly many variations in design, but all were characterized by a flat bottom made up of pine boards laid lengthwise, with battens nailed across to hold the bottom together. Oak frames, usually made from natural crooks, fastened the bottom to the pine planks that formed the sides of the vessel.

These craft were propelled by poles and oars, with a small sail used when the wind permitted. The Mohawk River batteaux built in Schenectady were apparently smaller and lighter than most, because of the shallow and often constricted channel they had to navigate and the several portages around which they had to be carried on their way west.

For the past several years the State Museum has been researching the contributions made by the Western Inland Lock Navigation Company (1792-1820) to the opening of westward transportation in New York. This private company was chartered by the New York Legislature in 1792 to find ways to improve the inland water route from Schenectady to the Great Lakes. Among the improvements made were the creation of several short canals, with dams and locks, that bypassed obstructions to navigation, thus issuing in an age of canal travel decades before the better known Erie Canal crossed the State from Albany to Buffalo.

The core of the Museum's research—known as "The Durham Project"—has been the discovery of archaeological remnants of these historic engineering works and the development of a plan for their preservation. Publications will result from this research when it is completed. In the meantime, this educational program is designed to bring some of the research findings to public attention during the bicentennial years of the building of the works which were the true beginnings of the Canal Age in New York.

In order to enhance public interpretation of this era of inland transport, a typical commercial batteau of the type built in the 1790s in Schenectady for the Mohawk/Oneida route has been recreated. Using archeological remains of earlier military batteaux in State Museum collections, historical accounts and drawings, and general technical data on small boats of the period, a probable design was determined. Construction duplicates original materials and techniques where possible.

GENESEE VALLEY CANAL

(Concluded from Page 7)

1876: Tolls collected were $14,700 vs. repair and operating costs of $23,000.
1877: NYS Legislature authorizes abandonment of the GVC.
1878: Last navigation season. Accumulated operating costs = $2.82 million. Revenues = $860,000. Net Loss = $1.96 million. This can be compared to the Erie Canal's cumulative net profit in 1888 when tolls were eliminated of $92 million.

LENGTH:	Miles	LOCKS:	No. of Locks
Roch. to Olean	107	Erie Canal Mud Lock	(1)
Dansville Side-cut	11	Genesee River Guard Lock At Mt. Morris	(1)
Extension--Olean to Millgrove	7	Roch. to Olean	104*
		Dansville Side-cut	8
Total	125	Oswayo Creek Guard Lock	(1)
		Allegany River Guard Lock	(1)
		Total	112 + (4)

*Locks #105 & 106 at Olean started but never completed.

1880: Canal right-of-way sold (with the exception of the Dansville side-cut sold to local farmers and two stretches near Rochester and Cuba Lake retained as Erie Canal feeders) to the Genesee Valley Railway Company for $11,400, less than $100/mile.
1883: The Genesee Valley Canal Railroad opened through from Rochester to Olean. Later became part of the Pennsylvania RR system.
1963/1971: Railroad abandoned to Wadsworth Junction/Rochester.

	Feet Above Sea Level		Vertical Feet
Erie Canal at Rochester	510		
		97 Locks	979
(Dansville Side-cut)		(8 Locks)	(83)
Summit Level (Black Creek, Oil Creek Reservoir, Ischua Feeder)	1489*		
		7 Locks	76
Allegany River at Millgrove	1413		
		Total Lockage	1055

*Highest of any U.S. canal.

SUMMIT 11.5 miles long between Locks 97 and 98.

ADIRONDACK WATERS: LIFEBLOOD OF CANALS

by David L. Kipp

New York is one of the few states to have still in operation a direct descendent of an inland waterway built during the canal-building craze that flourished in the second quarter of the 19th century. At that time, canals (called "internal improvements" along with roads and railroads) were seen as ways to link together regions of our then-expanding country, making it easier and cheaper for commerce to flow westward. For example, Philadelphia wanted to be connected with Pittsburgh on the Ohio River; the state of Ohio, to connect its namesake river with Lake Erie as did Indiana by way of the Wabash River; Illinois desired to link Lake Michigan with the Mississippi by way of the Illinois River; New Jersey, the Delaware River with New York Harbor; and Maryland, Chesapeake Bay with the Ohio River. These and other states caught the fever and built canals or canalized rivers to bring produce or raw materials from the hinterlands to manufacturing cities and ports of export in hopes of generating an economic boom.

New York's Erie Canal, loinking the Atlantic Ocean via the Hudson River to the Great Lakes at Buffalo, was the only one that in one continuous pathway truly pierced the Appalachian Plateau between the eastern seaboard and the interior midwest. It was the first completed (1825) among the major long-distance canals and was an instant and long-term success. Its tolls not only paid for its own construction, operation, and upkeep, but also subsidized construction and maintenance of a half dozen economically unsuccessful branch canals in New York and an Erie Canal enlargement to keep up with growing traffic, which did not peak until after the Civil War. By then, the competition from railroads was sending all but a few of America's canals into bankruptcy and abandonment. The railroads first lured away passenger traffic, then light freight, and were making inroads into heavy bulk cargos. Initial improvements made on

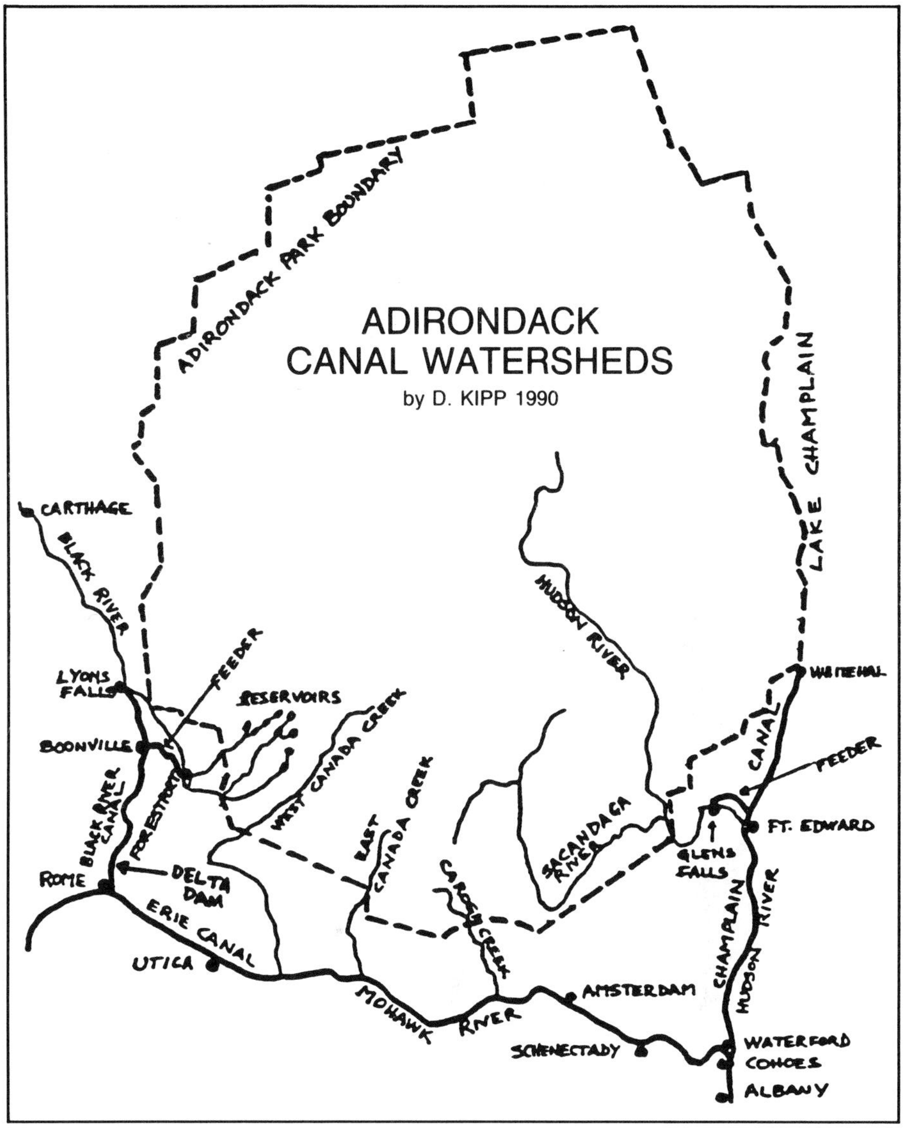

Black River Canal, Lock 64—top of the Boonville Gorge looking south. The canal here began a steep descent from the summit level down to the Mohawk River valley carrying water tapped from the Adirondack foothills ultimately to the Erie Canal at Rome.

the Erie Canal attempting to prevent defection of shippers to railroads were: doubling of locks and widening of aqueducts to prevent bottlenecks; and widening and deepening of the canal bed to allow bigger barges to be used. Later, locks were lengthened to permit two barges to be locked in tandem; mechanical changes to the locks and gates were tried to reduce the time of each lockage. However, as the 20th century approached, New York, now desperate to keep freight on the canal, eliminated tolls and began a second enlargement. But it was soon realized that this was only a Band-Aid and that what was really needed was an end-to-end modernization to complete the transition from the age of human and animal power to that of steam and electricity. Thus the New York Barge Canal system was born—constructed between 1903 and 1918. It remains little changed today.

Whether it was the original Erie (often called Clinton's Ditch for Governor DeWitt Clinton, its chief promoter), or the enlarged Erie which existed from the 1840s to 1918, or today's Barge Canal system, an adequate, dependable water supply was and is absolutely essential to canal operations. This is where the Adirondack watershed comes in.

A canal without an adequate water supply is like an automobile without gasoline—useless for its intended purpose. Erie Canal surveyors and engineers in their determination of the canal route considered not only the lay of the land but also water sources and feeders along the way, especially for the highest or summit levels.

In New York State, builders of the Erie and some of its branch canals were blessed by geography and topography with two major water sources that made the canal system possible: Lake Erie and the Adirondack Mountains. Lake Erie, which is over 500 feet higher than the Hudson River at Albany, supplies the western end of the canal with what would seem to be an infinite supply if it were not for the fact that it's not downhill all the way to Albany. After reaching a low point near Syracuse, it is necessary to surmount the divide between the Great Lakes and Hudson/Mohawk river watersheds in the vicinity of Rome, New York, before descending through the Mohawk valley to the Hudson River at Cohoes. It is to this summit level and the remaining section of the Erie Canal eastward that the Adirondacks supply their precious waters so vital to the Erie's success. The Adirondack streams supplying the Erie directly are East and West Canada creeks and Caroga Creek. In the past, the Black River and some of its tributaries also supplied water by way of the Black River Canal.

(Continued on Page 10)

ADIRONDACK WATERS: LIFEBLOOD OF CANALS

Delta Dam on the Mohawk River—a few miles north of Rome, this dam impounds the drainage from the western Adirondack foothills for today's Barge Canal. It replaced the old water delivery system by way of the now-abandoned Black River Canal.

(Continued from Page 9)

Two New York branch canals were also inextricably linked to Adirondack water for their operations—the Champlain and the Black River canals. Both received 100% of their water from the Adirondacks. The Champlain Canal lies in that ancient transportation corridor which links the Hudson River with Lake Champlain. The vital importance of this link to the north was highlighted by the French and Indian War and Revolutionary War battles that swirled along its route at Ticonderoga, Lake George, and Saratoga. It was recognized as a natural water transportation route in an early but unsuccessful attempt at a canal by the Northern Inland Lock and Navigation Company in the 1790s. So when the Canal Commission headed by DeWitt Clinton planned New York's original canal system, the Champlain was included as a branch canal to be constructed concurrently The Champlain was completed shortly before the Erie (1823) and its direct descendent continues today as part of the modern Barge Canal system. The Adirondack water it used came primarily from the Hudson where it was tapped above Glens Falls and directed eastward to the Champlain Canal in a 12-mile-long navigable feeder. Today, the Champlain uses the Hudson River itself as the canal from Albany to Ft. Edward and depends on the Glens Falls Feeder to supply its summit and the descent to Whitehall on Lake Champlain, thus continuing its dependence on the water released from the Adirondack Mountains.

Black River Canal

The Black River Canal is a bit of a different story. When the Erie Canal's early success brought boomtown status to such places as Buffalo, Rochester, Syracuse, Utica, and Schenectady, it generated considerable envy on the part of nearby districts where it was thought that a canal would also bring automatic economic prosperity. The people of the Black River valley, which lies on a north-south axis between the Adirondack foothills north of Rome and Watertown near Lake Ontario, were no exception. They petitioned Albany for many years for a canal link to the Erie and many surveys were taken and alternate routes considered. It was finally approved by the legislature in 1836 and construction began. The deciding factor for this authorization was not necessarily the persistence of the Black River settlers but the fact that the Erie Canal needed the water that a Black River Canal could deliver. Before the Black River Canal was completed, the Erie Canal would occasionally have to shut down in late summer and early fall when there was insufficient water to operate the Rome summit level. The Adirondacks, acting like a giant sponge, retain winter snow melt and spring rains, slowly releasing them at a steady rate throughout the summer and fall, supplementing the Erie's supply when linked by the Black River Canal.

The Rome summit level must have sufficient water supply to provide for lockages in both east and west directions (each lockful of water is eventually lost to the canal when it reaches the lowest point) and make up for losses. Even on the old Clinton's Ditch, each lockage required 125,000 gallons of water (for a typical lock of 15-foot width, 90-foot length, and a lift of 10 feet) and traffic often exceeded 100 boats per day passing a lock. This ongoing operational need plus additional losses due to surface evaporation, seepage, gate leakage, and overflow (some water must be allowed to flow over the waste weirs to maintain the canal at the proper level) meant that large volumes of water must continually be supplied to the summit level. Adirondack water from the Mohawk River supplemented by the Black River Canal was thus a key to maintaining canal operations.

Black River Canal Combine Lock—one of several sets of combines on this "mountain goat" canal which had 108 locks compressed into a mere 35 miles as it surmounted the divide between the Mohawk and Black River valleys. The sign reads, "Black River Canal—1837-1922—Connected Mohawk Valley with Black River country opening up 90 miles of navigable waterway."

The Black River Canal was begun in 1838, suffered through a stop-work period when New York State experienced a shortage of funds, and was finally completed in 1855. Its junction with the Erie Canal was in Rome. From there it paralleled the Mohawk River, branching off to follow the Lansing Kill up the Boonville Gorge across a tongue of the Tughill Plateau, arriving at its summit level at Boonville. It then descended into the Black River valley to meet the river at Lyons Falls. The river was by 1861 made navigable as far north as Carthage. The Black River Canal is justifiably called the "mountain goat" canal as it had 108 locks in 35 miles (3 locks per mile). Compare this to the original Erie's 83 locks in 363 miles (4.4 miles per lock). Many of the abandoned locks can still be seen along Route 46 between Rome and Boonville and along Route 12 north of Boonville.

Summit Level Supply

To supply the summit level of the Black River Canal, the Black River was dammed at Forestport and a 14-mile-long navigable feeder was dug from there to Boonville. This feeder can still be traversed today by canoe, and its towpath is an excellent hiking or cross-country skiing trail. Water in excess of the needs of the Black River Canal was allowed to flow into the Lansing Kill and from there to the Erie Canal by way of the Mohawk River. As traffic on the Erie increased during the middle of the 19th century, the Black River proved insufficient in dry years and several Adirondack lakes were dammed to make reservoirs that could release water when the Erie was in need. These bodies of water included Woodhull, Bisby, Twin, North and South Branch, Sand, Canachagala, and White lakes. This water supply route continued to function well into the 20th century when the Mohawk River was impounded by Delta Dam and Lake. These were built north of Rome to supply the much larger needs of the modern Barge Canal completed in 1918. By then the Black River Canal had already lost the battle with the railroads for shipping the region's lumber and ceased operations in 1925. It was because it was a channel for bringing Adirondack water to the Erie that it lasted as long as it did. Other New York branch canals like the Genessee Valley, Chemung, Crooked Lake, Chenango, and Onieda Lake canals were abandoned in the 1870s due to expenses far exceeding revenues.

It is interesting that Verplank Colvin, the renowned surveyor and cartographer of the Adirondacks, by virtue of his intimate knowledge of the headwaters of all the streams that exit the mountain massif, had strong ideas about the channeling of the mountains' vast water reserve. In his report to the legislature for 1874–79, he reiterates the canal system's dependency on the Adirondacks:

"In New York the great canals . . . are all of them, on their upper levels, dependent upon the Adirondack region for water supply; so that, but for the feeders from the streams of the wilderness, the canals themselves would be impracticable." [Verplank Colvin, Topographical Survey of the Adirondack Region: Report to the Legislature 1874–79 (Weed, Parson & Co., 1880), p. 329.] In an earlier report, Colvin promoted a remarkable visionary scheme for sending water from the high Adirondack lakes to the canal or waterway most in need (including the St.. Lawrence):

"The amount of water available from the wilderness for each of these canals (Erie, Black River, Champlain) is nearly equal, and may, by means of skillful engineering, be made extremely equitable in its distribution, in a manner which the peculiar character of the region renders possible, and which—though at first apparently astounding—is actually simple. The barometrical leveling of my survey in showing the altitudes of lakes has developed most important facts. It has been found that the interlocking headwaters of the rivers, flow often, for great distances in opposite directions and at inconsiderable differences of elevation; while the topography of the intervening country after examination of the defiles and ravines has been found to frequently admit of the construction of connecting canals; so that with aid of carefully arranged dams and reservoirs the water could be made to flow in either direction. Thus a single mind at the Capitol of the state, may control by telegraph the flow of numerous and considerable rivers; and at the tap of the key, turn so many million cubic feet of water from the St. Lawrence to the Mohawk, the Hudson, or to Lake Champlain! Such a proper system of reservoirs in this commanding position would prevent the stranding of boats in one canal for lack of that water which at the same instant might be swelling in floods through the affluents of the rivers feeding the others—by excess of rains on the distant watersheds—and would afford to each the quantity to which its commerce and its importance entitle it." [Topographical Survey of the Adirondack Region: Report to the Legislature 1873/74 (Weed, Parson & Co., 1874), pp. 104–05.]

Certainly, Colvin's reports emphasizing the rôle of Adirondack forests in supplying a steady flow of water to three of New York's canals were a factor when the 1885 law establishing the Adirondack Forest Preserve cited this as one of the reasons for maintaining the forests intact. Shippers who were in favor of the law may not have been so much devoted to the forests as interested in a healthy canal system that gave them a hedge against high railroad rates.

The fact that Colvin's schemes of interconnected, telegraph-controlled headwaters never came to fruition does not diminish the ongoing rôle of the Adirondack watershed as the supplier of the lifeblood for New York canals, past and present.

[David L. Kipp is an engineer with Eastman Kodak. He is a member of the A.C.S. and several regional canal societies. His writing has appeared previously in American Canals.*]*

Black River Feeder Canal—Through this channel, the waters of the Black River and an array of mountain reservoirs provided enough water to operate the Black River Canal and keep the Erie Canal's summit level watered in dry years. It carried the water from a dam at Forestport 14 miles to the summit at Boonville. It is still navigable today, at least for canoes, and the towpath is in great shape for hiking and cycling.

Black River Feeder Dam—Along with several upstream tributaries dammed as reservoirs, the water flowing off the shoulders of the southwest Adirondacks into the Black River drainage was here sent by a 14-mile-long navigable feeder to the summit level at Boonville. From there it descended through the Black River Canal joining the Erie Canal at Rome.

NATIONAL RECREATIONAL TRAILS FUND ACT

[The following exchange of letters will be of interest to A.C.S. members, especially those with memberships in other organizations which might also wish to endorse this proposed legislation.]

William E. Trout III, Ph.D., President
American Canal Society
Dear Dr. Trout:

This year, as Congress considers a major highway bill to pour tons of concrete and lay more asphalt, those who enjoy nature and the outdoors should be pleased to know that there is something in the bill for them, but it needs your support.

The "National Recreational Trails Fund Act" has been included in the Senate's omnibus transportation bill. Recommended in the 1987 Report of the President's Commission on Americans Outdoors, the Trails Fund Act mirrors the successful Aquatic Resources (Wallop-Breaux) Fund (which taps boat fuel taxes), creating instead a $50 to $75 million trust out of unrefunded "nonhighway recreational fuel" revenues. The trust's balance is dispersed annually to the states for recreational trail needs.

Trails are our access ways to nature, scenic beauty, wildlife and the great outdoors. And yet, a poorly maintained trail can damage vegetation, destroy wildlife habitat, increase soil erosion and add sediment to streams. The Trails Fund Act would help keep our trails in good condition, but the bill will only make it to the President's desk if it has the support of trail users nationwide.

That is why an endorsement from your organization is needed and would be much appreciated as a step toward making this legislation a reality.

Sincerely,
Steve Symms

The Hon. Steve Symms
United States Senate
Washington D.C. 20510

Dear Senator Symms:

As president of the American Canal Society, I want to offer our support to your Senate Bill 2150 to establish a National Recreational Trails Fund.

The American Canal Society, "Dedicated to historic canal research, preservation, restoration and parks" was established twenty years ago to serve as a medium of exchange between the nation's state and local canal societies and individuals concerned with the wise use of America's historic canal resources. There are thousands of miles of historic canals and inland waterways in this country, some dry, some watered, some urban, some rural, and there are many dedicated people working to use these historic assets to create parks, trails and open space. Canals and waterways, being linear, are ideal corridors for recreational trails of all kinds, from hiking and biking to canoeing, and along these canals are locks, aqueducts and other works as special points of interest—the Ancient Castles of the New World.

We would be glad to work with you and to assist the National Trails Advisory Board.

Sincerely yours,
William E.. Trout III

LIFE MEMBERS

Our official Life Member count now stands at 75. Check with Chas. Derr to find out how to join this special group.

ENCOUNTERING THE CHAMPLAIN

The EMITA II, which has recently been completely reconditioned, docks at the Skenesborough Museum in Whitehall, NY at the foot of Lake Champlain.

By Mary Cassai
(Photos by Bill Shank)

Everyone has heard of the Erie Canal—once irreverently dubbed "Clinton's Ditch" after the New York governor who spearheaded the drive to build it. But how many people have ever heard of the Champlain Canal? And of those who have heard it, how many know why it was built?

I was to discover both answers for myself during my first canal adventures ever during July of 1993. Doing the Champlain was a kind of afterthought. I had been asked by an editor to do an article on the Erie Canal. How could I do an article on the Erie, I reasoned, without cruising all 341 miles of it? So I promptly called the Mid-Lakes Navigation Company at Skaneateles, New York, to book a reservation on their canal boat, the *Emita II*. But the available bookings on the Company scheduled determined that I would have a week

As we left Troy, the following lined up for their photo: John Hulchanski of Latham, NY; Malcolm Horton of Schenectady; Lillian Alvord of Casenovia, NY; and Kay Stevens of Schenectady. All are members of the Canal Society of New York State and all but John are members of the Hudson Valley Chapter of the Steamship Historical Society of America. Lillian is a "Star-Customer" on the Emita II.

Mary Cassai (the author of this article) on the upper deck of the Emita II.

free between the two segments of the Erie trip. Rather than fly home to Chicago and back, I decided to use the July 5-7 Champlain Canal trip as a kind of "layover."

It was a lucky thing for me. The eighty-mile Champlain Canal has a charm all its own—its tempo, its lush, unspoiled woodlands, even the passengers on board—were in marked contrast to the Erie. The Champlain, I discovered, was born of a political move. Happily for us, De Witt Clinton and the New York State Legislature had to guarantee internal improvements to the people of the north country, to win their support for the Erie Canal. The prize was a canal of their own.

Without so much as a breath in between, I was picked up on Monday morning, the fifth of July, by crew member Chris Banks, and driven to the floating docks of Troy (an experience in itself: since the Atlantic Ocean tides sweep 150 miles north up the Hudson as far as Troy, the dock level rises and falls six feet between tides). Not always surefooted on a city sidewalk, I was a bit dubious about a floating dock. But the crew members handed me comfortably on board.

Thirty dedicated canallers were on the trip with me—members of the American Canal Society, some of whom love the Champlain trip so much they do it every year—and first timers like myself. Following a continental breakfast and an orientation lecture by Captain Dan Wiles, we were on our way through the Federal Lock at Troy and on to the north Hudson River.

What a difference from my Syracuse to Troy passage! I had watched the *Emita II* negotiating the twenty-two locks of the Erie, churning through those locks in the company of barges and pleasure craft bound for Fourth of July celebrations, and finally dropping an amazing 420 feet to sea level at the Hudson.

There was something wonderful about having truckers and drivers of the New York Thruway yelling and honking their horns at us—roaring by at 55 or 60 miles an hour—tickled to see us sailing along at a stress-free eight or ten miles an hour and waving wildly back at them. All the honking and hollering and waving created for us a kind of camaraderie among the great brotherhood of American travellers.

Now the *Emita II* eased her way like a lady of leisure along the peaceful stretches of the northern Hudson. A good stretch of the Champlain Canal uses the natural waterway of the Hudson River.

At Troy we found two of the Mid-Lakes boats tied up to the floating dock—the "Emita II" and the recently-built "Canvass White", one of a fleet of hire-boats patterned after the English "Narrow-Boats".

Gone were the historic towns of Mohawk Valley like Herkimer, Oriskany, Amsterdam and Little Falls. Instead of busy towns; miles of lush green forest lined the canal.

And of course there was some development along the canal—beautiful homes and boat launches and dock-divers so close you felt a part of them. I remember one occasion when a mother dog and her puppy ran down to meet us, not a bark out of them, tails wagging frantically. They urged their bodies so close to the edge of the canal I was afraid they might make a flying leap for our deck. But they never did—just ran alongside our boat to the end of some invisible territorial demarcation.

Just like the Erie, the Champlain boasts lovely locktenders' gardens—marigolds, begonias, petunias, and daylilies everywhere—and with beautifully maintained locks, bright with the blue and gold of New York State's official colors and bordered by landscaped picnic grounds. The locks are a feast for the eyes and a credit to New York State maintenance, and the lock-tenders.

It was at one of these locks that I felt most touched by a hand from the past.

I had heard over and over again about the old canal locks and the beams the lockmaster had to push against with all his bodily strength to open the gates and let the larges and packets enter the locks. I had visited the marvelous Erie Canal Museum in Syracuse, and then seen some of the beautiful old photographs and daguerrotypes of the lockmaster in action. I finally got a close-up look at one of the old locks.

It happened during a brief layover in Schuylerville. Dan Wiles, Captain of the *Emita II,* docked at this lovely spot in the Champlain Valley and offered us the opportunity to see an eighty-year-old canal powerhouse still operating beautifully—huge gears meshing with each other—works of art, really—and the powerful cables black and shiny with grease, and the water current flowing first one way then another as the boat was locked in and then released to the next level of the canal.

But as impressed as I was with the workmanship of the powerhouse, I discovered something that was for me even more striking. I happened to fall in step with Paul Mattice, an avid canaller and marvellous conversationalist, on a stretch of picnic ground some distance from our captain and his passenger audience. Suddenly, I became aware of something remarkable just ahead: a section of the 19th-Century Champlain Canal!

A typical lock on the Champlain Canal, as we entered from the south end. Several small boats were ahead of us, and we locked through together.

Captain Dan Wiles, with three of his crew-members.

The old canal was left exposed. A film of bracken coated the surface of the narrow waterway. Her eastern shore was carefully manicured and splashed here and there with brilliant tiger lillies and marigolds like the rest of the parkgrounds of the 20th-Century lock. But her western shore was lined with ancient trees that must have been tender shoots when the original canal was finished in 1823. Best of all, an old wooden lock gate had been left untouched. Its west side lay broken and in ruins, but the east gate was intact, the beam poised as if waiting for the hand of the lockmaster, silently calling for the master's little ones to ride the "wooden horse" as their papa braced his feet on the ridged towpath and then threw the full weight of his body against the beam, opening the lock to the next boat passing through.

To me that old lock was a thing of beauty. It told the story of the country's first great transportation system: a ditch dug only four feet deep and forty feet wide, with boats lazing through a mile and a half an hour, and passengers clustered on the roof listening for the cry: "Low Bridge!" It spoke of a

(Concluded on Page 14)

Here is the Fort Edwards lock where we left the Hudson River and began our journey overland, via the cut-canal, to the foot of Lake Champlain. After traveling the summit level, our last locks took us down-hill.

Who said there was no commercial traffic on the Erie Canal System? Here we passed an industrial barge and its tug-boat, carrying plate-glass and other commodities to a port along the upper Champlain.

ENCOUNTERING THE CHAMPLAIN

At the Schuylerville Lock, we docked the Emita II and Captain Dan Wiles explained the working of the lock machinery. Here Dan signals to the lock-tender to put the upper gates in motion.

(Concluded from Page 13)

young country opening up in every direction, and of the American "Can do!" spirit that believed itself capable of solving whatever problems lay ahead.

Nor are natural beauty and travel fellowship all the Champlain Canal has to offer. For the Champlain Valley witnessed some of the most dramatic moments of early American history. The Champlain Canal bore us past Bemis Heights, true site of the famous Battle of Saratoga and the surrender of British General Burgoyne. And as for American literary history—just imagine docking, as we did, at historic Fort Edward. Then imagine sailing into the pages of James Fenimore Cooper's *Last of the Mohicans*, with the action surrounding Fort Edward and Fort William Henry. Then imagine riding through the same territory where the French commander Montcalm defeated Colonel Munro; where the beautiful mulatto heroine Cora defied the evil Magua during the struggles of the French and Indian War. Luckily we spent the night, not at a fort, but at the elegant Landmark Motor Inn (Glens Falls) during the round trip to Whitehall, New York and back to Troy.

I had heard of Whitehall—a brochure on New York's new system of Urban Cultural Parks listed Whitehall as the "birthplace of the United States Navy." It's one thing to read this in print. It's quite another to sail into the storybook harbor of the real-life Whitehall, to walk down the *Emita's* gangplank and into the old Skenesborough Museum, then to sit down before a wonderful sound-and-light diorama production of the history of Whitehall and of the birth of the United States Navy.

It was the crisis point of the American Revolution. The British fleet was amassed to the north—and the colonials had no navy. So they commandeered the woodlands and resources, including the sawmill and forge, of an absent landowner and Tory named Philip Skene, and whipped up a navy inside of three months—just enough naval power to keep the British forces to the north from linking up with British armies to the south and strangling New York and New England.

With its many landmarks the Champlain Canal is indeed a passageway into colonial history, into Revolutionary War history, even into the literary history of great American classics. It is a passageway into a world of natural wilderness and unspoiled beauty. But for me, the heart of the Champlain Canal beats in the old wooden lock at Schuylerville.

At Schuylerville, we discovered one of the locks of the old Champlain Canal in fair state of preservation. One of the old wooden, hand-operated lock gates was still in place.

PENNSYLVANIA CANAL SOCIETY TOURS PAW-PAW TUNNEL

View of the west portal from inside the tunnel. Note the original wooden guardrail in the foreground, and the "rubbing rail" at the lower right, to prevent boats from coming in direct contact with the masonry tunnel wall.

by Bill Shank

Beautiful fall weather blessed all but about one hour of the Pennsylvania Canal Society fall tour of the west end of the Chesapeake and Ohio Canal, October 4 and 5, 1991. Driving into Cumberland, for most of us, involved a trip along the new Interstate 68 (former U.S. 40) and some gorgeous fall foliage in the mountains of western Maryland. One of the most impressive spectacles on this new route is the 380-foot-deep mountain cut at Sideling Hill, which can be seen for miles both east and west. On the north side of the cut is an interesting information center which features the geology of the area, so dramatically exposed by the cut.

Approximately 40 members of the P.C.S. from a three-state area attended the various sessions at the Holiday Inn in Cumberland Friday and Saturday evenings. Bob Keintz, Tour Director, showed a series of slides of the points to be visited on our tour. This was followed by David Johnson, of the C. & O. Canal Association, with more slides collected during his travels along the canal as a

(Concluded on Page 15)

PCS TOURS PAW-PAW TUNNEL

(Concluded from Page 14)

regular "level walker." The second evening, after an excellent buffet dinner, P.C.S. President Jim Oliver conducted a business meeting at which new directors were elected for the coming year, and plans for the spring meeting (May 1-2, 1992) along the Schuylkill and Union canals were presented. Following this, "Zip" Zimmerman took over the meeting and ran an interesting series of slides and movies of historic canals in Maryland and Pennsylvania on a large V.C.R. unit set up for this purpose.

Fair skies greeted us Saturday morning for the bus tour, which began at the Fifteen-Mile Creek Aqueduct on the canal. Here we were joined by

Fifteen-mile Creek Aqueduct, a beautifully preserved masonry structure—the first stop on the tour.

several C. & O. Canal Park Rangers, including John Frye, a long-time member of P.C.S., who guided us to the various campsites and restored artifacts along the canal towpath.

Fifteen-Mile Creek Aqueduct has been maintained in excellent condition by the Park Service. All the masonry has been carefully repointed to restore it virtually to its original condition. The same is true for most of the other C. & O. structures that we visited on the balance of the tour.

Our second stop was at the famous Paw Paw Tunnel. This 3,118-foot, brick-lined canal tunnel was originally planned to be built in two years, at an estimated cost of $33,500. Instead, it took 14 years (1836 to 1850) to build, and cost over $600,000, delaying the use of the full length of the C. & O. Canal until the canal had been almost superseded by the building of the B. & O. Railroad over the same route and beyond.

The walk from the parking area to the west portal of the tunnel was approximately 3/4 of a mile. We encountered a number of bicyclers on the towpath. Most of our group walked the full length of the tunnel, with the promise of a possible visit to several locks at the opposite end. Some hardy souls hiked *over* the mountain, returning through the tunnel. Your reporter was satisfied to investigate the west end of the tunnel, taking note of the burn marks of the towropes on the well-preserved wooden guard rail along the towpath inside. It was almost as though the canal boats had just stopped running a few months earlier.

The tunnel being wide enough for only one boat at a time, the canal company assigned watchmen to regulate traffic at both ends. The downstream boat was supposed to have the right-of-way. The story is told of two stubborn canal boat captains who met at the center of the tunnel, both refusing to back out. The tunnel was tied up for two days, until a company official built a roaring cornstalk fire at the upwind end of the tunnel and forced the offenders out! In the heydey of the canal, boats were frequently lined up at both ends waiting for clearance to enter.

One of the locks at Oldtown, with a log lockhouse at the left. There is no water in the canal at this point, but the rebuilt lock gates are still in good shape. Note the bypass flume to the right.

We ate a satisfying picnic lunch immediately after our tunnel tour, and our bus then took us to several restored locks near Town Creek Aqueduct, where the bus driver got stuck in a deep hole. We had plenty of time to enjoy a walk on a long rewatered section of the canal, until a heavy-duty towtruck arrived and got us going again.

Last stop of the day was at Oldtown, where there are several restored locks, a tollhouse built of logs, and a recently-constructed replica of a C. & O. Canal freight boat. A heavy rain began just as we were leaving, with your reporter catching the boat but almost missing the bus, after a very wet walk along the towpath! Back at the hotel, we all dried out and had good food and a pleasant evening.

The tour group preparing to enter the west portal of the Paw Paw Tunnel. The east end of the tunnel is visible at the center of the photo.

A watered section of the canal near Town Creek Aqueduct. The decrepit lock gates have been supplemented by a weir to hold back the water above the lock.

SOUTH CAROLINA CANAL PARK REOPENS

The Old Santee Canal State Park has reopened, almost two years after Hurricane Hugo forced its closure. The park, located at Moncks Corner, South Carolina, preserves the heritage of the Santee and Cooper Canal, which from 1800 to 1850 was the principal link between Charleston and its rice-producing hinterland. The park reopening coincides with the inauguration of a new Interpretive Center, featuring models of canal construction and operation, as well as an introduction to the natural history of the area.

CRUISING THE WATERWAYS BY CANOE

Off to Utica, following the Fourth of July celebration in Syracuse.

by Wilson Greatbatch

[Editor's note. Most of this material was originally published, in slightly different form, in the Clarence Bee, and appears here with the permission of the Bee Group Newspapers.]

In an attempt to experience what our forefathers experienced years ago, three generations of Greatbatchs spent two and a half weeks during the summer of 1984 rowing canoes along the 342-mile Erie Canal, from the Niagara River to the Hudson. The idea developed over a two-year period of planning by number-one son Warren and his uncle, Les Wright, formerly of Akron, New York, and now of Tucson, Arizona. Unfortunately, Uncle Les was incapacitated by a massive coronary during the spring and wasn't able to go. He participated only vicariously, through telephone calls and cards enroute, and by a slide show afterward. Warren developed the concept of rowing the canoes, procured the charts, planned out the stops, and tested out the whole thing between Tonawanda and Lockport before we left. To him goes the credit for making it all work.

Our entourage ranged from three to six people, and from two to four canoes. People joined in and dropped out as their time allowed. Warren rowed about 320 of the 340-mile stretch. I did about 300 miles. Warren's Uncle Dee (Dee Wright, of Akron) did about 200 miles. Other family members included granddaughter Ericka Greatbatch and number-three son Kenneth and his wife Sharon, from Keene, New Hampshire. Also with us for two days was Henry Kautz of Clarence.

Most of our canoes were fitted with oars, which were much more efficient than paddles. Warren had a beautiful rig, with a "Mad River" canoe, a sliding seat, and nine-foot, feathering sculling oars. I had a used Sears fiberglass wreck with a board seat and homemade outrigger, made from electrical conduit, for my seven-foot oars. Warren and I will never agree on which worked out better. Mine was more stable and shipped less water in the wake of a fast motorboat, but his was easier to row.

Start of the trip at Lockport — hopeful hoisting of a largely useless sail.

I told him I made up for that difference in pure brute strength. Granddaughter Ericka spent alternate days riding with Dad and with Grandpa. I asked her which boat was best. After considerable thought, she said, "Yours is more comfortable, but Dad's is faster." She'll make a fine diplomat some day.

There is a distinct disadvantage to traveling backwards: you can't see where you're going. After the first two days, I bought a rearview mirror. It helped, but my canoe is still covered with red streaks. I think I hit every buoy between here and Albany.

I cheated: I had a sail. Actually, the sail wasn't very useful, though. The winds on the canal were most capricious, and I found I could always row faster than I could sail. I didn't use the mainsail after Brockport (except for one instance in Syracuse—see below) and after Utica, I didn't even use the jib.

At night we camped by the side of the canal, or stayed in motels when the weather was bad. We were pleasantly surprised to find so many motels so close to the canal. Sometimes it was good to have a hot shower and dry out our gear after a day of rowing in cold rain. Many nights we set up our tents right on the well-kept lawns at the locks. The lockmasters encouraged us to stay there, since our presence discouraged vandalism after they closed the locks at 11 p.m. and went home.

Our trip was not without problems. Warren fell through a rotten board in one dock, and we had to take boards off to get him loose. I slipped on a muddy bank and fell, hitting my head. It was five minutes before the canal stopped going around and I could stand again. Fortunately, we got nothing worse than bruises. We broke one oar and two oarlocks, holed one canoe, and I had to row the last two days with one six-foot and one seven-foot oar.

On The Erie Canal

The lock people were very accommodating, letting us use their lavatories, and their shops when we needed repairs. They were very proud of their well-kept machinery and grounds, and each took great pleasure in showing them to us.

We overnighted in Baldwinsville, near Syracuse, on July 3rd. On the Fourth, we rowed the few remaining miles to Lake Onandaga, and pushed our canoes up on the shore at the county park that you can see from the Thruway. Our families from Clarence and from Schenectady drove there and we had a Fourth of July picnic in the park. This was one of the rare times I used my mainsail. It began to rain, and we hung the sail from the trees to make an awning over the picnic table. About mid-afternoon, our families waved us off with sparklers and we continued on our voyage.

The rowing wasn't as exhausting as I had feared it would be. We were dead tired by evening, but completely refreshed and ready to go in the morning. We usually covered 15 to 20 miles each day, depending on wind and current. From Syracuse to Utica, we had three inches of rain in three days. The Mohawk River was in flood and we had a four-mile-an-hour current pushing us, as well as a tail wind. The next-to-last day we did thirty-one miles, and the last day we did thirty-five.

After two and a half weeks, we pushed the canoes up on daughter Anne's back lawn. She and husband Larry Maciariello have a four-acre estate on the Mohawk River in Alplaus, just north of Schenectady.

What a tremendous feeling of accomplishment! We traveled the same route that Governor DeWitt Clinton did in 1825, when the canal was built, and

Son Warren and Granddaughter Ericka under way on the New York Barge Canal.

Ericka locking through — not in accordance with Corps of Engineers procedure.

we did it with our own backs and arms. We had problems, accidents, and rainstorms, but the important thing is, We Made It!

It was a trip I shall remember all my life. Never again will the Thruway be a boring six hours to Albany. Here's the spot where we never saw Herkimer from the canal because the canal banks were so high. There's the spot in the Montezuma Swamp, where the canal lies just beyond the hedgerow, where we saw all the great blue herons. Here's where we had the picnic. There's where I snapped an oarlock when I hit that red buoy. There's where the flooded feeder stream filled the canal with limbs and trees and we could hardly get through.

I often travel on business to the West Coast in four hours. In the canoes, after four days of travel we were still seeing Buffalo newspapers on the stands and hearing Buffalo stations on my pocket radio. Incredible!

It's hard to imagine that the old timers did all this with just men and horses and no engineering precedents to guide them. It makes you proud of your heritage, proud of New York State engineers and entrepreneurs who, a century and a half later, are still doing new things. I'm glad we went.

Three Years Later — On The Ohio River

Three years later, we undertook to top our previous feat by conquering the Ohio River. This time, two of our five-person team had health problems that precluded rowing, so we took along a 21-foot Wellcraft runabout, as a chase boat.

We trailered the runabout, and car-topped the canoes, to the Beaver River, which empties into the Ohio a little below Pittsburgh. I belong to the Buffalo Yacht Club, and we have reciprocal privileges with most yacht clubs around the country. This enabled us to take advantage of the hospitality of the Beaver Valley Yacht Club. We put our canoes in at their dock, launched the runabout at a nearby ramp, and parked our pickup truck and trailer in the yacht club yard for a couple of weeks. We had planned to start immediately on our trip, but the forecast was for thunderstorms, so we decided to wait until morning. Sure enough, down it came while we were eating dinner at a nearby restaurant. We spread our sleeping bags on the yacht club covered balcony that night. Nearly every day after that they forecast thunderstorms on the river, but it never rained once. Good thing!

For the next two weeks, we slept in everything from a tent on an island to picnic tables in a yacht club pavilion, and even a few nights in riverside motels. The hospitality of the yacht clubs was outstanding. In addition to the one at the Beaver River, we visited the East Liverpool (Ohio) Yacht Club, the Wierton (West Virginia) Yacht Club, the Martins Ferry (Ohio) Yacht Club, the Magnolia Yacht Club in New Martinsville, West Virginia (which turned out to be defunct, and converted into a bingo hall), and the Marietta (Ohio) Boat Club. Most of them charged us little or nothing to tie up for the night. Everyone seemed interested in our trip and our canoes. They were most kind.

At the mouth of the Beaver River, this is presumably where the beavers get the raw material for their dams.

The Ohio River locks were fantastic, after those on the Erie Canal. Some are 1,200 feet long and 150 feet wide. They can take one of the humongous pusher tugs with 12 barges (4 long by 3 wide) all at one time. My, those locks looked big from our little canoes!

Our eating habits varied. I usually tried to get on the river early and drift with the current while I ate breakfast, which was granola or some other dry cereal in powdered milk plus water. For lunch, sometimes our two canoes met with the runabout and we stretched our legs for a time. Other times, I just ate peanut butter and jelly sandwiches in the canoe. I carried breakfast and lunch makings in a foam cooler, along with quite a lot of fruit, until someone stole the cooler out of the canoe one night in Marietta.

Rowing — Hard Work

Rowing was hard work, and I carried a gallon of water along with a generous supply of salted pretzels to munch on during the day. I still managed to get pretty dehydrated, which gave me some urinary problems until I forced myself to drink the whole gallon of water each day.

We always stopped before 5:00 p.m. and either cooked a good supper or went to a restaurant. We had a complete camp kitchen with us and a 12 by 12 foot tent. Sometimes we set up the kitchen on shore, and other times we cooked on the deck of the runabout. We stressed flexibility rather than precise planning.

We had hoped to make many stops in the little towns along the way. However, we never saw most of them, because of a railroad or a steel mill between the town and the river. A few, like Wellsburg, West Virginia, Sistersville, Ohio, and Marietta have really developed their waterfronts, and made them attractive and accessible. In places like these, we stopped, looked around, and spent money. It occurred to me that the Ohio River might be going through something similar to what happened on the Erie Canal a couple of decades ago. There, the vanishing commercial traffic was gradually replaced by recreational traffic. Now every town on the Erie Canal has an attractive waterfront and good docks and facilities. They cater to the recreational boaters instead of walling themselves off behind a railroad or factory.

We usually made 20 to 25 miles per day, rowing. One thing we hadn't anticipated was the heat and humidity. It seemed that on most days the temperature went over 90°F. It made rowing so hard that we actually cut our trip short. This had been planned to be longer than the Erie Canal trip, but ended up only about 200 miles. The last day

(Concluded on Page 18)

The fleet's in — daughter Anne's backyard on the Mohawk near Schenectady.

CRUISING THE WATERWAYS BY CANOE

Three Rivers Stadium in Pittsburgh — here begins the Ohio River.

(Concluded from Page 17)

I rowed, it was 98°F and 90 percent humidity, and I rowed 17 miles into a head wind. We tried to get our rowing in early in the day, both to be in the cooler part of the day and to avoid the wind, which seemed to rise in the afternoon. Of course, by traveling southwest, we were bucking the prevailing wind.

Marietta was a most enjoyable town. We arrived there during their bicentennial celebration. This was the first town in the old Northwest Territory. They had loads of fireworks, special events, tours of the Delta Queen sternwheeler, that passed us on its way to Pittsburgh, and street concerts by a Dixieland band. It seems that the trumpeter owned a print shop on Main Street, so that's where they played.

Friendly Ohio Valley

I was impressed by the friendliness of everyone in the Ohio Valley. Everyone says Hello to everyone when they meet. Even little kids do it, whether they know you or not. Everyone is polite, considerate, and willing to talk anytime. A real uplift! But also, everyone will willingly give you directions on how to get where you are going, whether they know the way or not!

Our people decided to stop in Marietta a few days and fish. I'm not a fisherman, so I had them car-top my canoe 50 miles up the Muskingum River to Rokeby, Ohio. I put my canoe in just below the lock and spent the next three days cruising down the Muskingum alone. After the Ohio River locks, the Muskingum locks look like children's toys. They are less than 50 feet wide, and are all hand operated. The lockmaster walks around a turnstile a dozen times or so, first to open the valves and then to open the gates.

I spent my first night above the McConnelsville lock, had supper in town, and then walked over to the lock to check it out. A neighbor directed me to the lockmaster's house. My chart said they were open 9:00 a.m. to 6:00 p.m., but he said No, they were open from 8:00 a.m. to 5:00 p.m., except that the next day he was on garbage detail and would have to leave early. I asked When. He said 8:00 a.m.! But, if I was there at 8:00 he would lock me through. I was there at 7:45, and he did.

I got to the Stockport lock about 11:00 a.m. and found no one there. Another canoe came along with a crew that had friends in Stockport. They spent 1½ hours calling to try to find the lockmaster. We finally portaged their canoe around the lock and started to unload mine. Just then, a little outboard came tearing up. It was the lockmaster from Luke Chute lock, six miles downstream, who ran up on his day off when the grapevine relayed our problem. Evidently the Stockport lockmaster was also out on garbage detail! When we got to Luke Chute, there was our outboard driver again, who locked us through on his day off.

Next lock was Beverly. Again, no lockmaster. Finally, the lockmaster's young nephew borrowed a wrench from a local truck and got the lock open, and we went through.

I decided the reason that the service on the locks is so poor is the lack of traffic. The reason the traffic is so sparse is the poor service on the locks. Catch 22!

The other canoe carried a retired farmer and a physician (nuclear medicine) who were on a week-long canoe trip from Coshocton to Beverly. They talked to each other non-stop. The doctor said to me, "We're so happy to see some company. You're the first boat we've met all week." Then they positioned themselves 20 feet to one side and proceeded to talk my ear off. I tried to ignore them, but couldn't. In my distraction, I ran my canoe into a tree. I tried to fall back, but they did too. I tried to pull ahead, but with two paddlers, they were faster. Finally, I just resigned myself to them being there and sweated out the rest of the day. Fortunately, they pulled out at Beverly. Oh blessed peace!

While driving up to Rokeby, I had spotted motels at 20-mile intervals, making sure I could see the river from the motel. Unfortunately, that doesn't mean you can see the motel from the river. In Beverly, on the way down, I finally had to stop, tie up in a bush, climb a 30-foot bank, and walk along the road until I found someone to ask where the motel was. They asked where I left my car. I said I was in a canoe. They looked a little askance, told me where the motel was, and took in their children.

Canoes with oarlocks are not the only strange craft on the Muskingum.

I got to Lowell lock at 11:55 a.m. Surprise, the lockmaster was there! He looked at his watch, said he took lunch from 12:00 to 1:00, and I should come back at 1:00. So I did some shopping, got a milk shake, came back at 1:00, and went through. As I left, he shouted, "You know Devola lock is out of commission, don't you?" Shock, shock!

Oh well, Devola lock is only six miles above Marietta, so I called Warren to meet me there with the truck at 4:00 p.m. I guessed it pretty well and arrived at 3:45. He arrived at 3:55. Perfect timing, and we pulled out. Thus ends the Muskingum saga.

Blennerhasset Island

One interesting stop later on the Ohio was Blennerhasset Island, near Parkersburg, West Virginia. A British nobleman built an estate there in the early days of the Northwest Territory. Aaron Burr gathered an expeditionary force there to conquer Mexico for the United States, but without the U.S. government's knowledge or permission. History got all mixed up, and he was eventually (evidently unjustly) tried for treason, largely due to the personal animosity of President Jefferson. He was acquitted, but people don't seem to remember that.

It was interesting to walk around the island and imagine the goings on. The manor house has been restored, but someone noted that the famous flood of 1937 buried the whole island under 12 feet of water.

So, down to Gallipolis, Ohio. This was a town built by French immigrants (Gallipolis = Gallic City), and the end of our trip. It was a good trip, but the next one will be shorter, and cooler.

CANAL FILMS NEEDED

Old Santee Canal State Park, a 250 acre park in Moncks Corner, South Carolina, is located on the site of the first summit-level canal in America. Construction of an 11,000 square foot Interpretive Center is nearing completion. Located in the center will be exhibits dealing with the Santee Canal, plantation life along the canal, the natural history associated with the park, and other features of historical significance in the area. In addition to the exhibits, there will be two theatres, featuring many different shows. It is intended that nature films will be shown in one theatre and films on canals in the other.

The park management is seeking information on VHS, 16mm, and slides dealing with canals that could be purchased, rented, or borrowed for showing in the theatre. To date, an entire library of nature films has been acquired, but only two films on canals. Anyone possessng relevant information is asked to make contact with Mary Anderson Seabrook, Curator, Old Santee Canal State Park, 900 Stony Landing Road, Moncks Corner, South Carolina 29461 [Phone: 803 899 5200]

PROTECTING THE D&R CANAL PARK

by James C. Amon

Editor's note. A "news brief" item in our February 1991 issue cited the Courier-News of Bridgewater, New Jersey as the source of information that the future of the Delaware and Raritan Canal "is threatened by uncontrolled and seemingly uncontrollable pollution." James Amon felt that the original article was misleading and that our summary compounded the fallacy. While regretting whatever editorial boo-boos we may have committed, we rejoice at having motivated Mr. Amon to produce the following rejoinder.

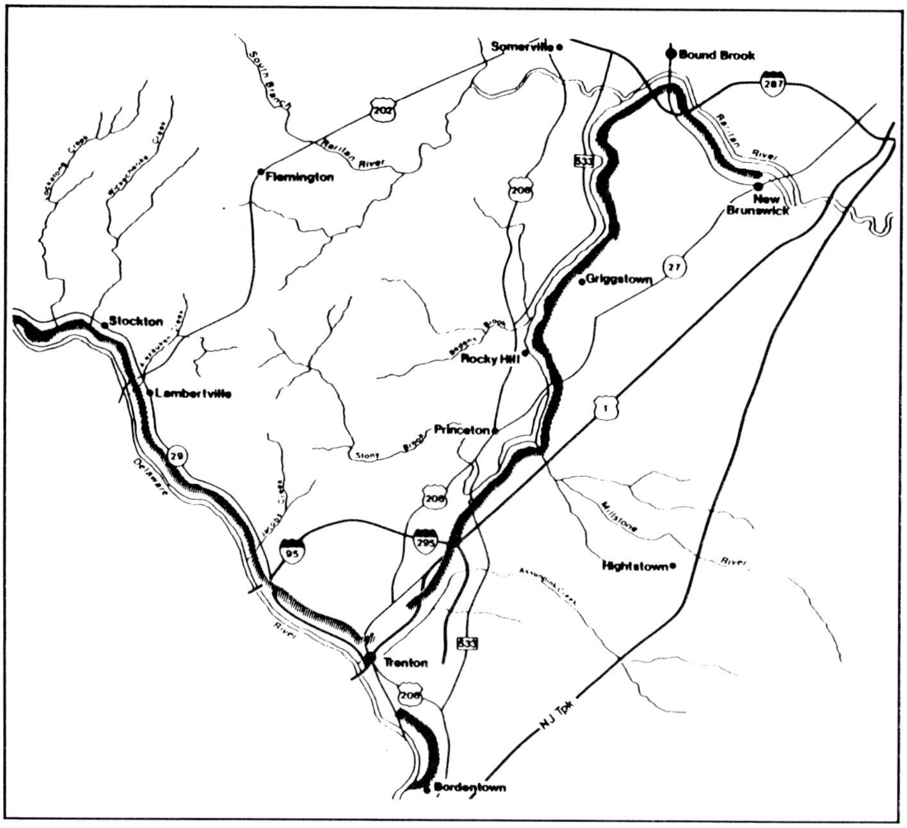

The Route of the Delaware and Raritan Canal

New Jersey's Delaware and Raritan Canal was converted in the 1940s to a water-supply system providing water for agriculture and business and as a portion of the drinking water for about one million present-day residents of the Garden State.

When the state legislature decided in 1974 also to make the historic canal (entered on the National Register in 1973) a state park, they realized that there was quite a tangle of government agencies making decisions that impact on the waterway. Their response—not as inane as it may seem at first—was to create a brand-new agency called the D&R Canal Commission. The commission was given the authority to review and approve governmental decisions that affect the canal: it oversees the State Park Service's actions to develop the park, the water supply office's actions to maintain the water flow, and the Department of Transportation's actions on roads next to the park and bridges over the canal, as well as relevant decisions by the four county governments and twenty-two municipal governments having jurisdiction over territory included in the canal park.

The canal commission was also given the authority to review private development that could have a harmful impact on the canal park, whether that impact were on water quality, flooding damage, or the park's visual, ecological, and historic qualities.

For eleven years, the D&R Canal Commission has been administering a land-use regulatory program aimed at protecting the canal and park. The commission has developed stormwater runoff standards for all new development, that are designed to reduce pollution and lessen the threat of flood damage within the 400-square-mile watershed of the canal. These regulations have been so effective that they have subsequently been adopted as statewide standards for New Jersey, and are in the process of becoming the standard for stormwater management recommended by the American Society of Civil Engineers.

The effectiveness of the regulations in protecting the aesthetic, ecological, and historic qualities of the canal park are a little harder to judge. The main element in this program is a requirement that projects be set back at specified distances from the park boundaries and that landscape buffers be planted between the projects and the park. Most of the D&R's supporters agree that these regulations have also been effective, with one or two exceptions. These exceptions, however, have motivated the commission to tighten the regulations and increase the level of protection for the D&R Canal Park.

The canal commission's protection program has not been able to do anything about conditions that existed prior to the creation of the canal park. Storm sewers from nearby streets, for example, were directed into the D&R from the time of its construction in 1834, and they bring pollution into the canal. The system also cannot prevent illegal dumping, a problem that parks everywhere in America must face.

The Delaware and Raritan Canal Commission has enjoyed the support of historical societies, environmental organizations, and the general public for its stringent program of protecting the D&R Canal. Developers may not be equally ardent supporters, but eleven years after the program was started, there still has not been a single lawsuit.

[James C. Amon is Executive Director of the Delaware and Raritan Canal Commission.]

Delaware and Raritan Canal State Park: Blackwell's Mills Bridgetender's House and Station. Photo by J. Amon, 1989.

CORRECTION

Notice in the November 1990 issue of *American Canals* of the publication of Frankel & Wiles' *New York State Canal Guide, Western Region,* failed to mention shipping cost. To purchase, send $19.95, plus N.Y. sales tax if applicable, *plus $2.00 shipping cost,* per copy, to Jeremy Frankel, P.O. Box 2414, Glens Falls, New York 12801.

JOINT NEW YORK AND PENNSYLVANIA CANAL SOCIETY FIELD TRIP A SUCCESS

By Bruce Russell - Contributing Editor

On the weekend of April 16 and 17, 1993, the Canal Societies of New York and Pennsylvania ran a two day Spring affair which included a Friday night orientation plus slide show on Pennsylvania's Lehigh Canal, a Saturday field trip using chartered busses to visit several important sites on the remains of the abandoned prism plus various other points primarily connected with the mining and movement of anthracite coal, a Saturday evening banquet and slide presentation, and an opportunity to look at additional canal oriented materials at the Canal Museum located in Easton, Pa. on Sunday. Officers of both canal societies put a great amount of effort into this weekend outing, and most participants learned a considerable amount of history on both the Lehigh Canal plus the region it served for about a century. Special mention must go to Albright "Zip" Zimmerman, Lance Metz, and John Miller for their running commentaries about the various points of interest during the tour.

Restored section of the Lehigh Canal At Walnutport. A weir has replaced the upper lock gates.

On Friday night most attendees viewed a slide presentation given by Lance Metz, who is no stranger to canal enthusiasts, historians, and industrial archeologists. He traced the development of the Lehigh Valley area from the mid-1700s when Moravian and German settlers began arriving, to the discovery of anthracite or hard coal which burns clean, to the initial surveys and eventual building of the Lehigh Canal and associated gravity railroad as a transport system for bringing the fuel out of the places where it was mined and to the major markets of New York City and Philadelphia. Lance explained that three artificial waterways met at Easton, Pa. Here a loaded boat or an empty could go in one of three directions. It was likewise explained that while the Morris and Delaware Canals were entirely man made channels or artificial rivers, the Lehigh Canal was a mixture of "cuts" or channels and so called "slack water navigation" in which a series of dams along the Lehigh River created artificial pools.

Also explained were the gravity railroads which constituted the earliest railroads in the nation. These lines, built during the late 1820s and early 1830s, were used to move coal from the mines to a point, usually at the foot of a mountain, adjacent to a canal. Here it was trans-shipped from railcars to canal boats for its onward journey to market. The gravity railroads used stationary steam engines to hoist strings of cars up mountains where they were then allowed to coast down the opposite side using gravity. They were both primitive and ingenious. Without them, the canal era would not have been possible since even the best canal engineers weren't able to design waterways to climb mountains.

Bill Shank, Publisher of *AMERICAN CANALS*, with Bruce Russell, Contributing Editor of *AMERICAN CANALS*.

Restored lock-house at Walnutport.

The Friday evening orientation held at the Easton Inn was a thorough presentation and provided a background to what would be viewed on the Saturday field trip.

Saturday dawned sunny but blustery, testimony that Spring had not fully arrived in this portion of Pennsylvania on the periphery of the Pocono Mountains. About 85 people boarded two chartered busses and the tour began. Each bus had two members of the respective canal societies to provide narration and commentary. As we headed northwest we paralleled the Lehigh River which empties into the Delaware at Easton. It was explained that at many locations the road we were traveling over was sitting on top of the former bed of the Lehigh Canal. However at other places this was not the case and the remains of the waterway were evident. Although worn considerably the prism of the Lehigh Canal seemed intact. Here and there crumbling locks were likewise visible. The canal was last used about 1935 although a few segments lingered until 1942. As a through route for bringing coal from the mines to the cities on the coast it was finished by the mid-1930s since normal railroads had long since captured its business except for local customers. When the Morris Canal from Philipsburg (opposite Easton, Pa.) quit in 1924 there was no longer any through water route to New York City unless the Delaware and Raritan Canal much further south was utilized. When the Delaware Division Canal from Easton to points south ended operations all possible routes were closed. Thus the Lehigh Canal and its associated waterways went out of service within a 15 year period, ending what had been a colorful period.

Our initial point of interest was Weissport, Pa. which was the location of major boat building and maintenance facilities. The Lehigh Coal & Navigation Company which actually owned the Lehigh

Restored coal-mine car in front of the Anthracite Coal Mine Museum in Lansford, Pa., originally owned by the Lehigh Coal and Navigation Company. (Russell photo)

The visitors look over the dry lock on the Lehigh just east of Freemansburg, with the partially restored lock house in the background. (Russell photo)

Visitors had a ride on the "Josiah White" canal boat at the Hugh Moore Park, on the lower Lehigh Canal.

Canal employed many local men to perform tasks associated with the waterway.

We then passed through the Lehigh River Gorge where a weigh lock of the Lehigh Canal was located. Entering Mauch Chunk, now called Jim Thorpe, Pa., the administrative building of the Lehigh Coal & Navigation Company was seen. Our tour came, however, not to admire old buildings but to take a look at what remains of the Switchback RR which transported coal from the mines across the mountains to the canal. We stopped our two busses where the sloping alignment of the railroad was still intact and plainly visible. A short section of narrow gauge track had been restored plus a replica freight car with seats for passengers. To thoroughly explore the right of way of this amazing 19th Century feat of engineering would consume two or three days. Nevertheless our brief stop afforded an opportunity to view something of this line. The Switchback gravity line remained operational until 1933, hauling tourists up and down the mountains. A visit of this location actually combined canal history with industrial archeology since both fields are involved.

The next place we came to was the town of Lansford, Pa. This was a community where many European immigrants lived and found work initially in the coal mines and on the gravity railroad and of course the canal. The group stopped at a restored chapel originally built about 1850 by Congregationalists. Abandoned and derelict for many years it has been put in good shape mainly by the work and dedication of volunteers.

Prior to lunch we spent about one hour at the Anthracite Coal Mining Museum located just outside Lansford. What had been the "wash shanty" for miners at a large facility has been turned into a series of exhibits on coal mining in the region.

Leaving the museum we headed to our lunch spot and enjoyed a buffet meal. After this we reboarded the busses and traveled parallel to the canal in the direction of Easton. We then stopped at Walnutport, Pa. where an 1828 vintage lock tender's house has been restored plus an adjoining lock chamber, minus working gates. Furthermore for approximately 4 miles the canal has water in it, thus making it look similar to the way it did prior to closure in the late 1930s. In the early 1950s a man by the name of Earl Snyder began to look into bringing back to life a small portion of the defunct Lehigh Canal so future generations could appreciate the great era of towpath canaling. The Walnutport Canal Association was created and in 1985 they purchased the diminutive stone lock tender's house and set about restoring it. The Lehigh Coal & Navigation Company maintained houses for its lock tenders as a kind of fringe benefit and of course to have them on site to work the locks.

After the Lehigh Canal ceased operation, the lock tenders still on the Lehigh Coal & Navigation Company's payroll were allowed to purchase the tiny homes where they lived and worked. Frank Kelchner, the last tender at Walnutport, did just this and he and his descendants remained until 1965 when others acquired it. At some point a small addition was added. Presently the Walnutport Canal Association owns and maintains the house as a museum open to the public. Our group had a chance to go inside the building and get a first hand glimpse of what it was like to live in the 1840s. Needless to say there was no central heating and toilet facilities weren't situated inside the home! This restoration is an example of just how much can be accomplished by dedicated volunteers.

The tour then passed through the village of Treichlers where many canal employees lived who weren't lock tenders. Canal captains made adequate incomes and many chose this hamlet to build homes and raise their families. In addition to anthracite coal, cement was also a major commodity carried on the Lehigh Canal. Several of the cement companies owned and maintained their own boats. After cement the next most common cargo was iron ore bound for the furnaces in and around Bethlehem. Some of the ore originated in New Jersey on the Morris Canal. It headed west to Philippsburg and then crossed the river to Easton where it was taken north on the Lehigh Canal to Bethlehem. During the first 25 years of its existence the Lehigh Canal also carried people in packet boats but this business was immediately lost to the railroads once the first rails were spiked down. Train travel was simply faster.

The next stop during the latter part of the day occurred at Freemansburg, Pa. where lock number 44 of the Lehigh Canal and its adjacent lockhouse remain in a semi-state of restoration. This house, like the one at Walnutport, dates from the 1820s and is of stone construction. An original structure, it serves as a living relic from a bygone

(Concluded on Page 22)

A Russian Orthodox Church in downtown Lansford, representing the wide range of immigrants who came here to work the mines, canals and railroads. (Russell photo)

PENNSYLVANIA CANAL TRIP

Walking along the abandoned right-of-way of the Switchback Railroad near Jim Thorpe. (Russell photo)

(Concluded from Page 21)

era. It is now listed on the National Register of Historic Places. At Freemansburg were also stables for mules and coal yards served by the Lehigh Canal. A group called the Old Freemansberg Association has charge of this historic site. The tour remained here for about 20 minutes exploring the entire premises.

The final stop of the day was at Hugh Moore Park where about 5 miles of the canal has been restored to its original appearance. During the spring, summer, and fall a pleasant ride is available aboard a replica packet boat pulled by a pair of mules. Called the JOSIAH WHITE after one of the principle figures of the Lehigh Canal, it takes passengers for a half hour trip up the waterway. Presently a new vessel is being built which will replace the 20 year old JOSIAH WHITE which is slowly deteriorating. It was a fitting way to conclude an excellent day's outing.

Later in the evening a banquet was held and the completed goals of the two canal societies were recognized. Members were able to reflect on the accomplishments made in the area of canal education, history, and preservation. America's towpath canal epoch was an exciting chapter in the nation's history.

A replica of one of the passenger-carrying cars on the Switchback, after it became a tourist attraction in the early 1900's. (Russell photo)

Left to right: Clarence Stephenson, a past Vice President of the Pennsylvania Canal Society; Bill Dzombak, Editor of the PCS "Canal Currents"; "Genie" Walton; and Denver Walton, Editor of the "American Canals" Bulletin. They are standing on a canal bridge to the Walnutport Lock House.

FLOATING ELEVATOR TRANSFERRING GRAIN FROM A CANAL-BOAT TO AN OCEAN STEAMER

By Bill Trout

This drawing of a floating grain elevator in New York City was drawn by C.A. Vanderhoof for "A Day on the Docks," an article by Charles H. Farnham in the May, 1879 *Scribner's Monthly*. The drawing is from an original copy of the article sent to us by Bob Trbovich. The text on page 36 describes the scene:

"Ocean steamers occupy several consecutive piers below Tenth street and give their locality a distinctive character. I went on board one of the ships and looked down her hold, four stories deep . . . . [One] hold was like an ant's hill — swarming with men loading sacks of grain. An elevator alongside raised the grain from a canal-boat, and then sent it down a spout pouring two six-inch streams of wheat. A man opened and closed the valves of the spout while two adjusted the sacks under it; four sewed up the bags as fast as they were filled at the rate of about eight per minute."

The article also mentions unloading coal and cabbages from canal boats, and along South Street "There are also canal-boats stowed away here in a slip for the winter, while the captain and his family live in the cabin and enjoy a season of metropolitan life." And if you think we live in an age full of criminals, read this article to see what it was like on the New York docks in 1879. Sometimes the police got their man, as in this story told by one of the officers:

"Sometimes the thief saws part of a plank out from under a pile of goods and then fills his boat; or he bores an auger-hole up till he taps a barrel of liquor, and runs it into his own cask. They have a thousand tricks of the kind. One of us lately heard a man working at a pile of fruit. He soon filled his boat under a hole in the planking and then started away; but just as he came out from under the pier, the officer above dropped a slip-noose over his head and hauled him up to the dock, and that thief was about as much surprised as any man you ever saw."

Floating elevator transferring grain from a canal boat to an ocean-going steamer.

THE TRANS-JERSEY SHIP CANAL—A DREAM UNFULFILLED

by Bruce J. Russell, Contributing Editor

At the beginning of the present century, most of America's old 19th century towpath canals had either gone out of business or were in the process of doing so, victims of technological change in the form of faster and more efficient railroad competition. Nevertheless, there were those who clung to the belief that transportation of goods by water was inherently more economical than rail transport, provided that travel time was not considered vitally important. Bulk commodities such as grain, cement, coal, soda ash, bricks, oil, and even lumber could be moved just as easily by water provided that navigable rivers and canals existed large enough to accommodate vessels in excess of 300 feet. If not bona fide ships, these vessels might be barges commodious enough to handle the equivalent of several railroad freight cars.

Along the eastern seaboard, there were plans to create an intracoastal waterway which would extend from the New York City area as far south as Key West, Florida. Using existing bays and rivers connected by a series of man-made canals and channels, it would be a protected and sheltered route for East Coast shipping. In other words, vessels would not have to venture into the Atlantic Ocean, but instead would travel on the western side of the 1,200-mile-long chain of sand bars and barrier islands which form a series of resort communities stretching from Point Pleasant, New Jersey to the southern tip of Florida, and include Atlantic City and Cape May, New Jersey, Rehoboth Beach, Delaware and Ocean City, Maryland, Virginia Beach, Virginia, and many others too numerous to mention. In order to gain access to them from the mainland it is necessary to travel over bridges or causeways which cross a bay, an inlet, or an estuary of some type. The planners of the proposed intracoastal waterway believed that if these bodies of water could be joined, an efficient coastal route for shipping might be developed. Goods produced in New York City could be loaded onto vessels and transported down the East Coast without the danger of shipwreck or foundering in the open Atlantic, or in time of war, being sunk by enemy submarines. The scheme envisaged a number of ports developing along these bays, complete with machinery for loading and unloading bulk cargoes. Furthermore, the proposed intracoastal shipping route would provide some measure of competition for the railroads which, by the late 1800s, were often controlled by those unscrupulous financiers known as "robber barons."

Army C.E. Surveys

By 1905, the location of the various canals and channels which would link the natural bodies of water had largely been determined. The U.S. Army Corps of Engineers as well as other federal government agencies began doing surveys up and down the coast to determine the specific details of the waterways which would be built using the latest types of dredging and excavating equipment. At the time, New York State was building its barge canal system from Albany to Buffalo as a replacement for the 1850-era Erie towpath canal, and observers were sent to learn first hand the techniques being employed on that ambitious project.

In 1907, an organization appropriately called the "Atlantic Deeper Waterways Association" was formed with the purpose of promoting use of waterways to move the nation's bulk freight and, in certain instances, passengers. Totally in favor

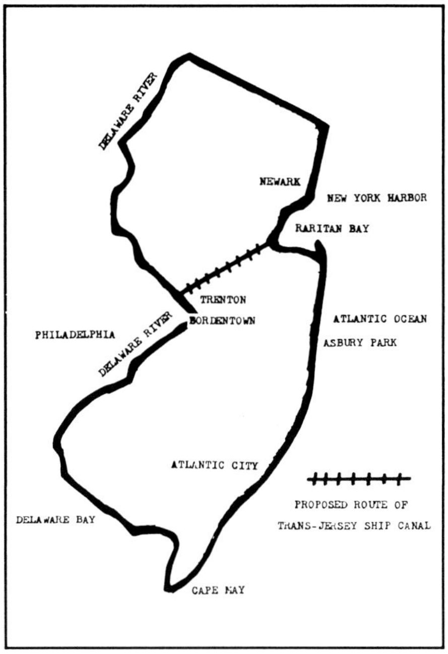

of the creation of an intracoastal waterway from the New York area to South Florida, the group on its own began to survey various routes where it felt canals could economically be excavated.

In certain instances, their proposals differed from those of the Corps of Engineers and other government planning agencies such as the Department of the Interior. For example, in New Jersey the approved route consisted of passage through Barnegat Bay, Manahawkin Bay, Little Egg Harbor Bay, Townsend Sound, et cetera, as far as Cape May, the extreme southern tip of the Garden State. At this point, the waterway crossed Delaware Day using a buoy-marked shipping channel which took it to Delaware City, Delaware. Here vessels were supposed to enter an enlarged and deepened Chesapeake & Delaware Canal providing access to Chesapeake Bay and ports such as Annapolis, Baltimore, and Washington, D.C. They would never have to enter the Atlantic Ocean

WHAT MIGHT HAVE BEEN. Had the Trans-Jersey Ship Canal been built, it would have been possible to have excursion boats circumnavigating lower New Jersey. Vessels such as this could leave New York City, enter the canal at Raritan Bay, and proceed across central New Jersey to the Delaware River. From here, they could travel to Delaware Bay, go around Cape May, and Return to New York via Atlantic City, Asbury Park, and Sandy Hook.

except for a short stretch between Sandy Hook and Manasquan where an inlet led directly into Barnegat Bay and the beginning of the intracoastal waterway. The Atlantic Deeper Waterways Association, however, rather than accept the idea of skirting New Jersey's coast before crossing Delaware Bay at almost its widest point, felt that a more advantageous scheme might be to cut across the Garden State at its narrowest point. In other words, they wanted to create a waterway which would connect Raritan Bay, which is less than five miles from New York's Borough of Staten Island, with the Delaware River at Bordentown, a small community situated a few miles south of Trenton, the New Jersey state capital.

An existing towpath canal, the Delaware & Raritan, had been dug in the 1830s, and when it opened for business was considered the best thing that ever happened for commerce in New Jersey. Goods which had previously been shipped around Cape May could now be diverted onto a safer, protected inland route. No longer would valuable cargoes be lost in shipwrecks. Of course, one drawback was that goods had to be transferred from the holds of sailing ships to the much smaller canal boats. The waterways group proposed nothing less than a replacement of the Delaware & Raritan Canal with a modern waterway capable of handling large ships and barges, in the process eliminating the need to transfer cargoes. They were aware of plans for widening, deepening, and converting to sea level the 19th century Chesapeake & Delaware Canal, in order that it might be used by ocean-going freighters, and believed that the same thing could be done to the Delaware & Raritan Canal in New Jersey. They saw no logic in having a ship leave New York harbor, travel all the way south to Cape May, and then proceed north through Delaware Bay to reach the enlarged Chesapeake & Delaware Canal. Their position was, "Why not take the most direct route and create a waterway for ships to enter Raritan Bay, proceed west toward New Brunswick, and then cross New Jersey at its narrowest point until the Delaware River is reached." Miles of circuitous navigation around Cape May would be avoided. In their opinion, an enlarged, deepened, and widened Delaware & Raritan Canal or a completely new waterway roughly parallel to it was a natural counterpart to a modernized Chesapeake & Delaware. Together the two canals would furnish the shortest, most direct route for ships traveling between New York City and Washington, with stopovers possible in Philadelphia, Baltimore, and Annapolis. Some of their inspiration undoubtedly came from Europe, where an ambitious canal-building program was underway as the 20th century dawned.

The Trans-Jersey Ship Canal, as the scheme came to be known, gained early support from both business and industry. Both viewed the waterway as a practical means of bringing commerce to the Garden State. Raw materials could be cheaply carried to industries, and finished products sent out, at a fraction of the cost of land-based transportation. They wrote of how a century earlier men of foresight and vision had built the original Delaware & Raritan Canal, and that their spirit would guide the creation of a new 20th century version. Further support came in the form of the federal Rivers and Harbors Act of 1909, which called for the creation of an inland, intracoastal waterway from Boston to North Carolina and later as far south as Florida. This piece of legislation was the result of two years of lobbying by the Atlantic Deeper Waterways Association.

(To Be Continued)

THE TRANS-JERSEY SHIP CANAL - PART II

If the proposed Cross-Jersey Canal had been built at sea level, as suggested in one plan, the route might have looked like this. Actually this is a view of the Cape Cod Canal where the Sagamore Bridge crosses it.

by Bruce J. Russell, Contributing Editor

Survey work for a ship canal across New Jersey began in 1908, and by 1909 the approximate route had been determined. Like the original 1830-era Delaware & Raritan, it would begin at Sayreville on Raritan Bay near New Brunswick. Ships departing from New York City could travel through the Kill Van Kull and the Arthur Kill, and finally enter Raritan Bay in the vicinity of South Amboy. Such a route would of course avoid the need to travel upon the open sea. From Sayreville, the actual entrance of the planned waterway, ships would head 31 miles in a southwesterly direction toward either Trenton or Bordentown on the Delaware River. Once on that river, which separates New Jersey from Pennsylvania, they would proceed south, passing Philadelphia and Wilmington, Delaware, before reaching Delaware City, entrance to the Chesapeake & Delaware Canal. Ship owners looked forward to creation of such an inland passage, which, by eliminating miles of circuitous travel around Cape May, would save thousands of gallons of fuel or tons of coal since the latter was still being used to fire ships' boilers.

The Trans-Jersey Ship Canal would have been 31 or 32 miles in length. One scenario envisaged excavating a sea-level waterway the entire distance, since the land was essentially flat with no natural obstacles such as mountain ranges or deep valleys. If constructed in this manner, it would have been almost identical to the modernized Chesapeake & Delaware Canal. Another plan involved building a flight of three locks at each end, giving a total lift of 56 feet. Yet another engineering consultant advocated combining the two three-lock flights into single chambers with a 56-foot lift. Instead of having six individual locks to build and maintain, there would be only two. Either way, the technology for doing this was certainly available in 1905-10. Engineers working on the New York State Barge Canal were in the process of designing gigantic lift locks to be installed at Lockport, New York, as replacements for the 1850-era flight up the Niagara escarpment. And in far-away Panama, the massive locks at Gatun, Pedro Miguel, and Miraflores were in the final planning stage. Lack of technology was never considered an obstacle to constructing the Trans-Jersey Ship Canal. Gigantic steam shovels and other types of earth-moving equipment were available as was the procedure for pouring concrete to create massive lock chambers. The actual transportation of excavated earth and stone would have been done using narrow gauge railroads, much the way it occurred in Panama.

Water to supply the planned waterway was supposed to come from the Delaware River, and would be delivered to the main channel by means of strategically placed feeder canals. In fact, a portion of the existing 1830-vintage Delaware and Raritan Canal was to be incorporated into its successor's feeder system. Naturally, this would insure its survival as a purely recreational waterway, rather than one utilized for commercial purposes. In a similar manner, a section of the 1850-era Erie Canal serves as a feeder to the 1909-1918-vintage New York State Barge Canal in the vicinity of Utica, New York.

DINING OUT AT COLONEL BALDWIN'S

The Baldwin Mansion, located at Baldwin Landing on the Middlesex Canal in Woburn, Massachusetts, was originally built in 1661. It was occupied during the late 18th and early 19th centuries by Loammi Baldwin, a great grandson of the original proprietor, a colonel in the Continental Army during the War of Independence, and a civil engineer who served as Superintendent of Construction for the Middlesex Canal. The building has now been renovated and opened to the public as Baldwin's Restaurant. Agents of the A.C.S. and the Middlesex Canal Association report that canal history buffs with gourmet tastes will be well served here at moderate cost. [Information provided by the Middlesex Canal Association.]

In 1908, most of New Jersey, including the corridor through which the Trans-Jersey Ship Canal was to be built, was still farmland or undeveloped tracts. Land acquisition costs would have been minimal, and little if any disruption would have been created by the excavation. Building supplies such as concrete, stone blocks, clay, et cetera, could have been delivered to the various construction sites via the existing Delaware & Raritan Canal, which by this time was using small steam-powered tugboats in place of the traditional mules to haul boatloads of cargo. In his book, Canal Boatman—My Life on Upstate Waterways, Richard Garrity describes how the still-operational 1850-vintage Erie Canal was used to bring construction material to places where the replacement barge canal was being excavated. Similar arrangements would no doubt have been made in New Jersey.

Unfortunately, the concept of a wide and deep canal across New Jersey able to handle ocean-going and coastal ships did not please the powerful railroad interests. From the beginning, they viewed it as a competitor, and began to wage an intensive lobbying effort to prevent it from ever being built. Furthermore, they were opposed to any type of inland waterway development in the United States, viewing it as a throwback to the long-gone canal era. In New York State, the New York Central Railroad, still under the iron grip of the Vanderbilt family, had waged a bitter battle to stop the barge canal system from being built. For reasons not entirely clear, they lost this fight and the canals were constructed. Until the 1960s, when interstate highways, the St. Lawrence Seaway, and a general decline of industry in the Northeast robbed it of commercial traffic, the New York barge canal system was a busy waterway network. From Albany to Buffalo and from Syracuse to Lake Ontario, it witnessed a steady stream of large barges filled with grain, coal, oil, salt, and other commodities. Likewise, in the Midwest the railroads used their political clout to keep public money from being spent on improvements to the Mississippi and Ohio river networks.

Opposition in New Jersey

In New Jersey, opposition came not from the New York Central but from the even larger and more powerful Pennsylvania Railroad. Beginning in the 1850's, both this multistate carrier and its arch rival, the Reading Railroad, began acquiring the Keystone State's canals. By this time, the canals were starting to suffer serious losses on account of railroad competition, and purchasing them wasn't difficult. Although they continued in use for a few more decades under railroad ownership, very little money was invested in them except for normal maintenance. Across the Delaware River in New Jersey, the directors of the Pennsylvania Railroad decided to acquire the then 45-year-old Delaware & Raritan Canal, which was experiencing financial problems caused by parallel railroads capturing much of its business. One of their motivations was to eliminate this waterway as a potential competitor. They knew it offered a shortcut between the New York harbor area and the port of Philadelphia. While they never conceived of the 1870-vintage Delaware & Raritan Canal as a serious rival for their freight trains, they must have known that a rebuilt Delaware & Raritan or a new waterway parallel to it and capable of handling the largest freighters then in use would constitute a major challenge. The directors of the Pennsylvania Railroad were aware that a sea-level canal had just been dug at Suez in Egypt, enabling ships from Europe to avoid the trip around the cape of Good Hope on their journey to India and the Orient. Furthermore, they knew that the technology existed to excavate the planned Panama Canal across the narrow isthmus separating North and South America. Consequently, they perceived it to be in their best interest to gain control of the Delaware & Raritan Canal and either retain it in its existing form, perhaps for pleasure boating, or simply shut it down after a few years. In practical terms, this was their best insurance against a modernized version coming into being.

It thus came as no surprise that in 1907 the directors of the Pennsylvania Railroad became apprehensive when organizations such as the Atlantic Deeper Waterways Association began calling for the building of a Trans-Jersey Ship Canal. Almost immediately, their lobbyists both in Trenton, New Jersey, and in Washington, D.C., went to work to make certain that no state or federal funding would be made available for such a scheme. The same tactics were used 50 years later by the railroad industry in its effort to scuttle the St. Lawrence Seaway. Unfortunately for them, by this time they had lost much of their former political clout, and the seaway opened in 1959 amidst much fanfare. But in 1907, the railroads were at the top of the heap in terms of wielding influence, and no matter what arguments backers of the ship canal across New Jersey could muster, the scheme was destined to fail. Without all-important government funding, its chances of being built were nil.

In spite of setbacks in New Jersey, work on the intracoastal waterway continued, with many canals and channels dug up and down the East Coast. After World War I, the scheme to convert the Chesapeake & Delaware Canal across the northern portion of the Delmarva Peninsula began to move forward, and by 1926, work had commenced. But in the Garden State, government money began being spent on highways and airports, and few people gave much thought to canals. In 1924, the hundred-year-old Morris Canal, with its incredible system of inclined planes, was abandoned with virtually nothing saved as a reminder to future generations of early 19th century engineering expertise.

In the late 1920s, there were plans to construct a Trans-Jersey Ship Canal entirely at sea level, eliminating the need for costly locks. Those in favor of building it pointed to the just-reconstructed Chesapeake & Delaware Canal, which was now handling ocean-going and coastal shipping between Philadelphia and Baltimore. They viewed a rebuilt Delaware & Raritan Canal as a logical counterpart to it, and sent observers to gather data. Perhaps on account of pressure from the New Jersey canal supporters, the U.S. Army Corps of Engineers did conduct studies of a modernized ship canal across the Garden State, doing surveys and cost analyses. A detailed, working model was built. The final recommendation of the Corps of Engineers was for a canal consisting of a 500-foot-wide channel with one set of locks at each end. The total rise would be only 10 feet, and the canal would be lined with riprap in some places and with stone retaining walls in others. An elaborate system of reservoirs fed by the Raritan River would have kept the summit level supplied with water. One of the reasons why the final proposal was for a lock canal rather than one at sea level was to satisfy the objections of those who believed that a sea-level canal would allow water from Raritan Bay to flow into the upper portion of the Delaware River. The rise in salinity might, so the argument went, kill fish and otherwise disturb a delicate ecological system.

In spite of the work done by the corps, the canal wasn't built, not even as a Works Progress Administration (W.P.A.) project to employ people during the Great Depression. Instead, in 1931 the existing Delaware & Raritan Canal was abandoned by its owner, the Pennsylvania Railroad. No longer could small boats, yachts, and other craft cross New Jersey at its narrowest point, avoiding the hazardous trip around Cape May.

Had the Trans-Jersey Ship Canal been built during the early years of this century, it would probably be used both for pleasure and for commercial boating. Two- or three-day cruises using vessels up to 400 feet in length, similar to those seen on the Rhine and the Danube, could easily make circle trips around New Jersey and perhaps go all the way to Washington via the Chesapeake & Delaware Canal. Leaving the foot of Manhattan Island, such ships, possessing staterooms, dining saloons, sun decks, and swimming pools, might call at Bordentown, Philadelphia, Cape May, Atlantic City, and Point Pleasant. In addition, if the Trans-Jersey Ship Canal had been constructed, it is certain that by now it would be lined with marinas, boat-building yards, and waterfront restaurants. Unfortunately, such scenarios, so appealing to romantics, ship buffs, and canal aficionados never came to pass. An opportunity was lost. But in this age of monster trucks and dependency on foreign oil supplies, it is interesting to realize that alternatives were once considered.

This interesting sketch of a so-called "Luggage Boat" sailing down the Merrimack River was part of a small brochure sent us by ACS Vice President Bill Gerber. These boats were also referred to as "scows", propelled variously by poling, by oar, by sail and by mules or horses on a towpath. They traveled between the town of Concord, New Hampshire and Boston, Massachusetts via the Merrimack River navigation system and the Middlesex Canal. Their special construction permitted them to travel in very shallow water.

BOOK REVIEW

Erie Canal Museum, *Photos from the Collection* (Syracuse: 1990).

Reviewed by Jeremy G. Frankel

Photos from the Collection is a compilation of 250 photographs from 600 kept on file for researchers and historians. The museum's total holdings number nearly 17,000, according to the preface. The book is divided into five sections: Life on the Canal; Canal Country; From Weighlocks to Waste Weirs; Canal Ports; and 20th Century Canal.

I have mixed feelings about this book. On the one hand, if the aim is to introduce someone to the world of canals, boats, canal life, and the architecture, then the book is a success. I imagine that this must be the market the museum had in mind. However, for the more serious researcher or historian, there are several errors and miscaptions, and the index certainly needs fine-tuning.

The book could almost be subtitled "The Effect of the Erie Canal on Syracuse," as half of the views pertain to Syracuse and its environs. Several of the captions exhibit errors, the most blatant probably being the one on page 69, photo number 217, where lock number 2 on the Waterford flight is credited to the Champlain Canal instead of the Erie Barge Canal. The vast number of people standing on the lockside suggests that the picture may have been taken at the official opening of this flight, in 1918. The Champlain Canal side cut can be seen in the photograph, and it is mentioned but not identified as such. A previous view, photo number 44 on page 17, which shows a barge entering a lock, is merely captioned, "A lock on the Champlain Canal, late 1880s." Close inspection of the footbridge abutment reveals the wording "STATE CUT," and comparing this view with number 217 shows that the lock is in fact number 3, the top lock on the side cut to the Hudson River. Hence, a more specific caption could have been applied to photo number 44.

In other instances, the captions are misleading. An example of this occurs on page 25, where photo number 67 depicts a view of the Erie Canal at Little Falls, just west of locks 37 through 39. The caption states that at Little Falls, the Mohawk River, the canal, and the railroad all had to squeeze through a narrow pass. Yes, this is perfectly true, but this photograph does not show that place. On page 18, photo number 48, instead of asking us to squint at an almost invisible badge on the locktender's hat, why not describe the function of the hand-operated winch in the foreground?

The worst news for the canal historian is the index, which is littered with errors and marred by omissions. There are places mentioned in the captions, such as Lakeville, Weedsport, and Schenectady, which do not appear in the index. A lock indexed under the Black River Canal and shown in photo number 113 on page 39, is in fact situated on the Champlain Canal. It might also have been useful to mention in the text that this lock (number 10) at Schuylerville, is a guard lock where the canal meets the Hudson River.

As was mentioned at the beginning of this review, this book is probably aimed at the novice, though the canal buff will also enjoy the many interesting photographs. It's a shame the book has been diminished in scholarly value by what appears to be poor caption research in identifying the scenes and by errors in the indexing; otherwise, it is an excellent value for the money. The book can be purchased from the Erie Canal Museum, 318 Erie Boulevard East, Syracuse, New York 13202, at a price of $9.95 (there is no mention of any charge for tax or for shipping and handling).

MULE POWER FORCE ANALYSIS

by William Dzombak

A canal boat moves forward because a force FF, acting on the boat, pulls the boat in the direction shown in Figure 1. That force is applied to the boat by means of a towrope tied to the front end of the boat, as shown in Figure 2; the other end of the rope is attached to the harness of a mule, or horse, that pulls on the rope and so moves the boat through the water.

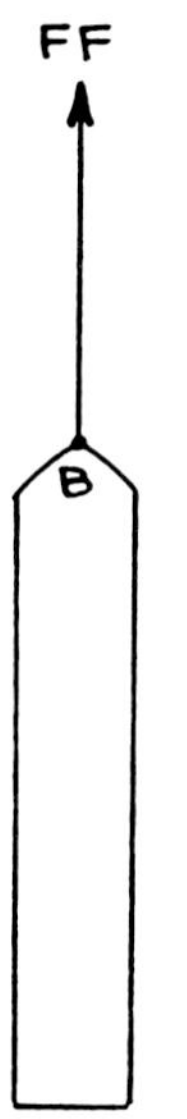

FIG. 1

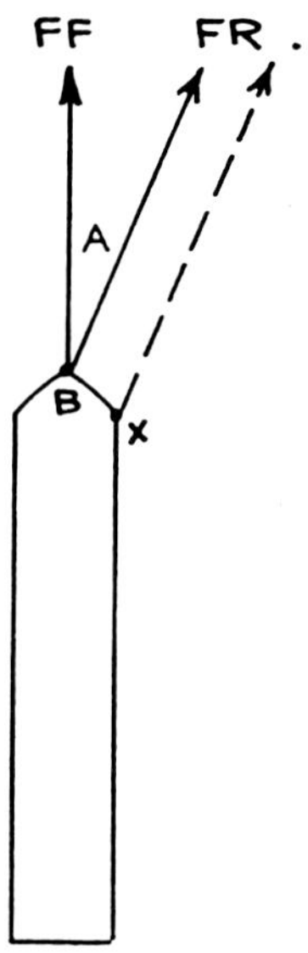

FIG. 2

FF = force acting to move the boat forward
FR = force exerted on the rope by the mule

Note: Each force on such a force diagram is represented by an arrow. The orientation of each arrow indicates the direction in which that force acts. The length of each arrow is a measure of the magnitude of that force.

The towrope defines an angle (A) away from the line of motion of the boat because the animal(s) pulling on the rope cannot walk on the water in front of the boat, but must be located on a towpath that runs parallel to the canal. To simplify the discussion and illustrations that follow, it is assumed that the towrope is attached as shown in Figure 2, on the centerline of the boat at the bow, and not at a point such as X, which was used on many boats. Later in this article, I will consider why the towrope was attached at point X rather than at point B.

When a horse pulls a wagon, all of the energy expended (work done) by the horse is delivered, without reduction or loss, to the wagon, because the horse walks on the line of motion of the wagon. On the other hand, when a mule, walking a towpath, pulls a boat, the animal and the vehicle (boat) do not move on the same line; the mule "pulls at an angle," so to speak, and therefore pulls the boat sideways, toward the towpath, as well as forward, parallel to the towpath. (See Figure 3.) The single force, FR, actually applied to the rope, could be replaced by two other forces, FF and FS, each acting separately, to achieve the same effects: forward motion of the boat at the same time it is drifting sideways, just as an airplane drifts off course in a crosswind, or if it is fitted with only one propeller.

Dixie and Daisy appear to be simply mules, but they are actually the latent foci of invisible but potent force vectors. [Photo by David T. Boyer for the Canal Museum at Hugh Moore Park.]

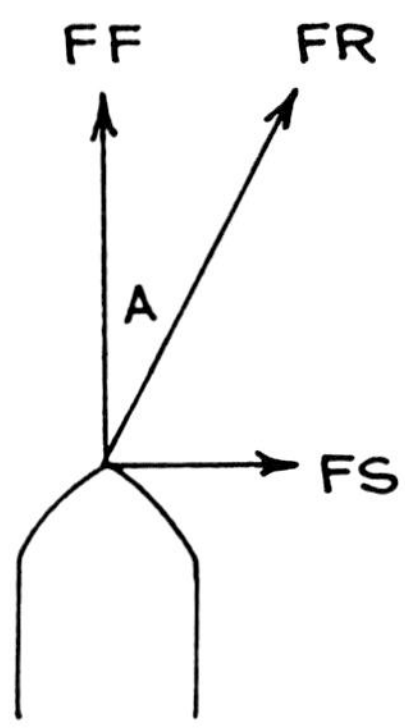

FIG. 3

Because the mule must walk on a line that does not coincide with the line of travel of the boat, the mule exerts two forces (pulls) on the boat: one pull in the direction of motion of the boat, and a second pull sideways, toward the towpath and the bank of the canal. The mule must do more than pull the boat forward; the mule must also do the added and unavoidable but useless work of pulling the boat sideways, which then requires that the boat be fitted with a rudder that can be used to keep the nose of the boat on a line that parallels the towpath. The extra work that the mule must do is therefore required to overcome the resistance to motion, or drag, created by keeping the boat rudder at an angle, and not amidships, on the axis of the boat.

Because the mule must walk on a line that is displaced to one side of the line of motion of the boat, the mule must do more work than would be required to pull a wagon of the same weight, even assuming identifical friction. A simple calculation can be made to discover how much extra work must be done by a mule pulling a canal boat instead of a wagon of the same weight and friction.

The force applied to the boat by the mule has been resolved into two component forces.

FR = force applied to the rope by the mule
FF = force that moves the boat forward
FS = force that tends to move the boat sideways
A = angle between towrope and axis of boat
FF = FR x cosine (A) = forward force
FS = FR x sine (A) = sideways force

The two forces can be obtained as soon as the angle (A) is known. To calculate that angle, it is necessary to make some assumptions regarding the distance between mule and boat and between boat and towpath, as shown in Figure 4.

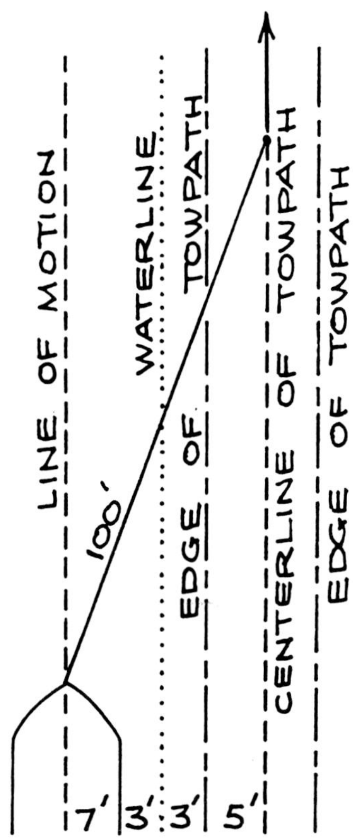

FIG. 4

Assumptions.

width of boat = 14 feet

minimum distance between boat and waterline on bank of canal = 3 feet

horizontal distance between waterline and edge of towpath = 3 feet

width of towpath = 10 feet

length of rope = 100 feet

Given those distances, the two component forces can be calculated by use of the relation, stated earlier, that the length of each arrow on a force diagram is proportional to the magnitude of that force:

FS = FR x sine (A) = FR x 18′/100′ = FR x 0.18

The cosine of that angle can be looked up in a table, or the length of the side of the triangle calculated (see Figure 5), to yield this result:

FF = FR x cosine (A) = FR x 98′/100′ = FR x 0.98

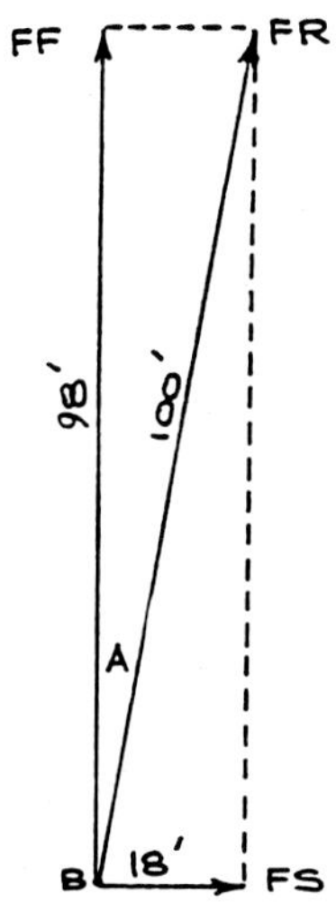

FIG. 5

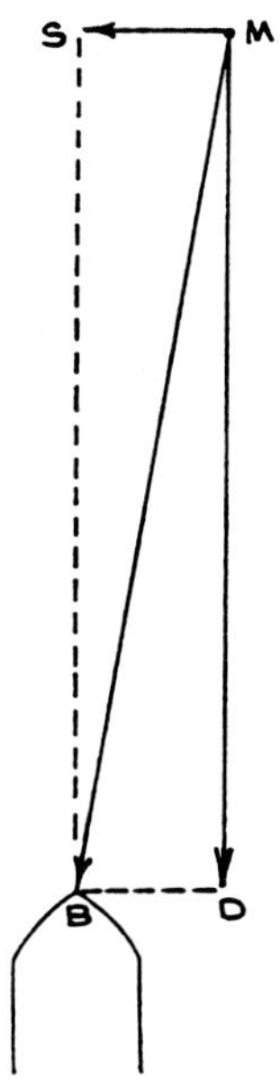

FIG. 6

To illustrate the interpretation of the two results just obtained, let us assume that the mule exerts a pull of 100 pounds on the rope.

FR = 100 lbs.

FS = FR x 0.18 = 100 lbs. x 0.18 = 18 lbs.

FF = FR x 0.98 = 100 lbs. X 0.98 = 98 lbs.

One calculation tells us that a reduction of force of only 2 percent (from 100 lbs. down to 98 lbs.) is the penalty paid by displacing the motive power (mule) off the line of motion of the boat and onto a towpath, the centerline of which is offset 18 feet from the centerline of the boat. The other result tells us that a rather large sideways force (18 lbs.) is unavoidably applied to the boat because the towline force is applied at an angle (A) to the centerline of the boat. The sideways force (FS) would be zero if the angle (A) were zero—that is, if the mule could walk in front of the boat, on the line of travel of the boat.

Having considered the forces involved in moving the boat, let us now examine the forces at work on the mule. As was shown in Figure 3, the mule applies three forces to the boat; the boat, in turn, reacts by applying three forces to the mule:

(1) counter-tension force

(2) toppling force

(3) towpath-tilt force.

In addition, three other forces that affect the mule wil be considered:

(4) immobilizing force

(5) towline-weight force

(6) skidding force.

Each of those forces will now be examined.

Counter-tension force (see Figure 6, MB)

When a mule leans into its harness, the limp towrope snaps taut and the boat begins to move in response to the force applied to the boat by the mule. At the same time, the boat exerts an equally large counter pull upon the mule. The two forces (pulls) are equal in magnitude but oppositely directed.

When the mule pulls on the towrope, the boat is pulled through the water; the pointed bow of the boat merely reduces, but does not entirely eliminate, the resistance to motion associated with pushing aside the water that stands in front of the boat. The drag on the boat, caused by plowing aside the water, produces a counter tension, or drag force, in the rope—the pull exerted by the mule being equal to the drag force generated by movement of the boat through the water.

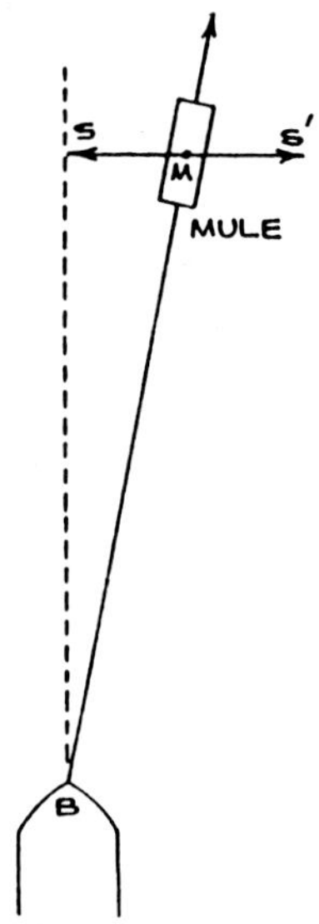

FIG. 7

The effect of that counter-tension force is to cause the mule to push off from its left legs and so to lean the right shoulder into the harness, in that way to exert a force that is in line with (opposed to) the counter tension in the rope associated with the inertia of the boat and the displacement of water by the bow of the boat. The mule quickly aligns himself with that counter tension by "squaring his shoulders" to the line of the rope, a position maintained by turning his body slightly so that he heads in a diagonal direction on the towpath, as indicated in Figure 7, that diagonal heading forming the angle (A) with the line of motion of the boat and the centerline of the towpath.

One of the components of the drag force is the force MS, indicated in Figure 6, and here labeled "toppling force." That toppling force pulls the mule sideways, toward the canal, and tends to topple the mule into the water, as indicated in Figure 8.

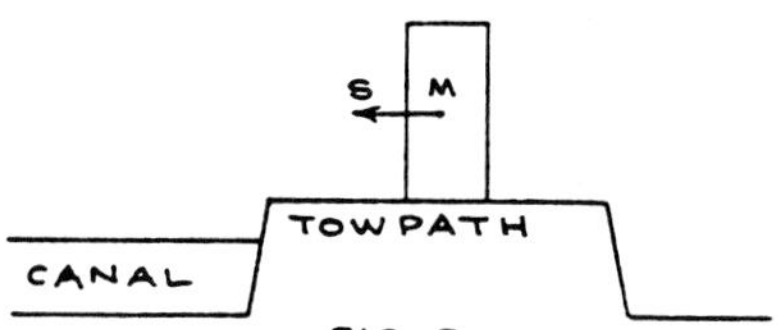

FIG. 8

Toppling force

When the mule applies a sideways force to the boat, the boat rudder is set to apply to the mule an equal but oppositely directed force that cancels the sideward drift of the boat and so keeps it moving parallel to the towpath. That rudder force is transmitted through the towrope to the mule, and acts to pull the mule sideways, toward the canal.

MB = counter-tension (drag force) exerted upon the mule by the boat

MS = sideways component of the drag force that tends to topple the mule from the towpath and into the canal

To counteract that toppling force, the mule changes stride: it leans its right shoulder into the harness and begins to walk diagonally across the towpath (Figure 7) and not on a line parallel to the line of motion of the boat, as had been depicted before (Figure 4). By walking a slightly diagonal path, the mule is able to exert a pull directed along a line that is an extension of the line of the rope; by walking that path, and so directing his pull along the line of the rope, the mule is able to apply a component of force (MS′) that equals and so cancels the force (MS) that tends to topple the mule. A mule walking a diagonal path may lean his right shoulder into the harness, but action of the boat rudder continuously pulls the mule back onto the centerline of the towpath during each stride the mule takes in the diagonal direction. That is, the mule's body leans slightly away from the canal while the mule's hooves tread steadily along the centerline of the towpath.

By thus compensating for the toppling force, the mule is able to distribute the pulling load equally upon all four legs, and not primarily upon the left-side legs. It would not be possible for a mule to walk a line parallel to the line of motion of the boat without leaning away from the canal. If the mule did not lean away from the canal, the uncompensated toppling force would cause the mule to place most of the pulling effort on only two legs, those nearest to the canal. Reflex action immediately compensates for that imbalance. The mule leans his body away from the canal just as we would do if harnessed to a boat.

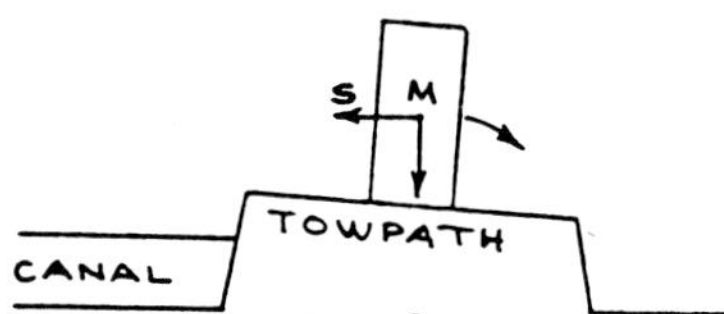

FIG. 9

Towpath-tilt force

The mule would have been assisted, also, if the towpath were inclined slightly, as shown in Figure 9, so that a mule straining to walk a diagonal line would be walking slightly downhill and so secure better purchase for the hooves. Some engineers specified that the towpath be tilted, if only to direct runoff toward a drainage ditch located on the side of the towpath away from the canal. Tilting the towpath would have secured an additional benefit.

If the towpath surface had been tilted slightly, so that it sloped away from the canal, that slope would have forced the mule (and people standing on the towpath) to lean slightly away from the canal. That reflex leaning would subject the mule to the action of the force of gravity, which would tend to topple the mule away from the canal, and so cancel, to some extent, the toppling force (Figure 6, force MS) that tended to topple the mule into the canal.

The center of gravity of the mule (see the plumb line, Figure 9) would still fall within the "base" of the mule—within the rectangle defined by the span of his hind legs and his front legs—so that the mule would always be in a stable posture, though braced to counteract the toppling force. A tilted

(Continued on Page 28)

MULE POWER FORCE ANALYSIS

(Continued from Page 27)

towpath would have allowed a mule to walk a diagonal path that would have made it easier for the mule to distribute more of the pulling load onto his front legs, and in that way assist the beast to apply maximum pull, because all four legs would then be equally in contact with the towpath (toepath!) at all times. By walking a diagonal line on a tilted towpath, a mule would be "falling" slightly into the harness at all times, and thus the inert weight of the animal could have been pressed onto the harness to reduce the amount of muscular effort required to pull the boat along.

Speaking of inertia, when a boat at rest was to be started, the mule had to exert maximum effort to set all of that load into motion; after the boat had come up to speed (four miles an hour, say), the mule was able to pull more easily, because then the only work required was that needed to overcome the drag associated with the rudder torque and the parting (displacement) of viscous water by the plowing action of the bow of the boat.

No attempt will be made, here, to calculate the amount of work required to set into motion a canal boat weighing perhaps 30 tons, for that calculation would force us to think about the caloric value of oats, eventually. Practically, when the tow was downbound, the poor mule was often assisted by locktenders who were willing to ignore the regulations and "flash" motionless boats out of the lock by opening the gates before the water in the lock had fallen to the level of the water in the canal below the lock. By riding the surge of water that flooded out of a lock, a boat quickly acquired momentum and so relieved the mule of the most difficult part of his job, which was a tug-of-war between a one-ton mule and a 30-ton boat. The mule exploited the advantage of a toehold and thus prevailed.

Immobilizing force

Another force that would act on the mule, depending on the inclination of the towrope, is MK in Figure 10.

MB = drag force (counter-tension) exerted by the mule by resistance of water pushed aside by the boat

MK = component force exerted on knees of mule

Because the towpath is above the water level in the canal, and because the mule stands upon that elevated plane, the towrope attached to the mule would necessarily be inclined, as shown, and not horizontal. Because the towrope is inclined, there is a component (MK) of the drag force (MB) that acts to pull the body of the mule down closer to the towpath. A heavy stone, slung under the belly of the mule, or a rider astride the mule, would have the same effect—to exert pressure on the mule's knees. The mule would then walk with difficulty—like walking in thick mud that is ankle deep. If that downward force were large enough, the mule would be immobilized—unable to lift his hooves from the towpath.

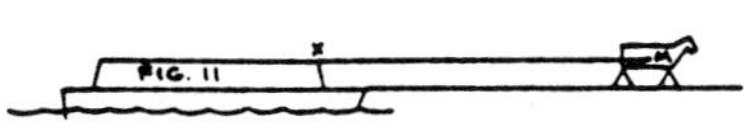

These mules (there are actually two in the photo, one well concealed behind the other) are illustrating several important principles of mule power force analysis, but seem more interested in human companionship. [Photo by David T. Boyer for Hugh Moore Canal Museum, Easton, Pennsylvania.]

To reduce the magnitude of that downward force, and so relieve the mule of that added burden, it was the practice to attach the towline to the boat at the highest point possible, as shown in Figure 11. That setting brought the towrope closer to the optimum position—horizontal. That consideration is probably one reason why a canal boat towline was never attached to the bow.

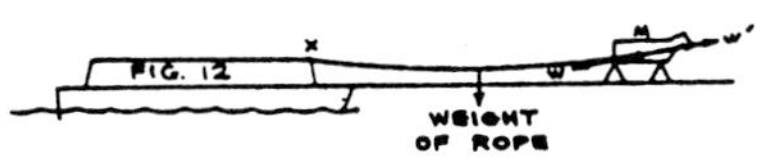

Towline-weight force

Another force acting on the mule is MW, Figure 12. The origin of that force is the weight (W) of the rope, which was supported by the two points of attachment, the boat and the mule. The portion of that weight supported by the mule is represented on Figure 12 as the force MW applied to the mule; to counter that rearward force, the mule leaned into the harness and so exerted an equally large force indicated as MW'.

The last item that needs to be addressed is the question of where to attach the towrope to the boat. In practice, the towrope was attached to the highest point on the boat, for the reason already discussed (immobilization force). The location of that point is marked as X on Figure 8 and on Figure 13. The rope could have been attached to the bow of the boat, at point B, but then a mast would have been needed to elevate the fastening and so raise the towrope into a horizontal position. Rather than encumber the design and construction of the boat with an appurtenance that improved performance of mule and boat only slightly, the designers provided for the towrope to be attached to a strong structural element that was already built into the boat—the frame of the fore cabin.

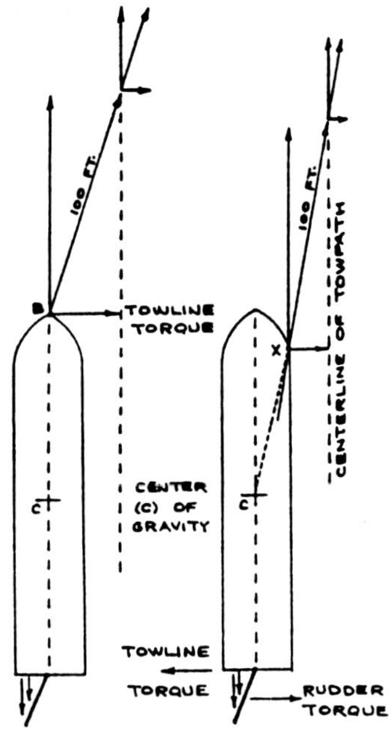

FIG. 13

It makes a difference whether the towrope is attached at point X rather than at point B. The sideways force is smaller when the towrope is attached at point X than it is when the rope is attached at point B. If in each arrangement the mule is on the centerline of the towpath and exerting the same pull on a line of the same length, the angle between the rope and the line of motion parallel to the towpath is less when the rope is attached at point X than when it is attached at point B. Because point X is 7 feet closer to the towpath than point B is, the sideways component of force acting on the boat is less than it is when the rope is attached at B. That is, when the rope is attached at X, the mule is 7 feet closer to the optimum pulling position, which would be directly in front of the boat.

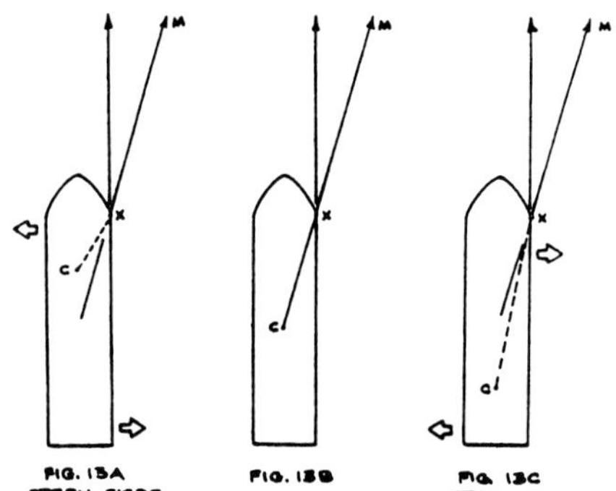

FIG. 13A STERN SKIDS TOWARD TOWPATH

FIG. 13B NO SKIDDING

FIG. 13C STERN SKIDS AWAY FROM TOWPATH

If the rope is attached at X, yet another force (a "skidding force") acts on the boat; to compensate for skidding of the stern (and bow, in the opposite direction), rudder force must be adjusted, and so the load on the mule is changed. The amount of rudder force required to cancel skidding, and so to keep the boat moving parallel to the towpath, depends on where the center of mass of the boat is located—whether forward (Figure 13A), or aft (Figure 13C), or on the line of the rope (Figure 13B). The location of the center of mass depends on how cargo was distributed when the boat was loaded. The center of mass is the point (C) where all of the weight of the boat can be imagined to be concentrated.

As is shown in Figure 13B, the towrope force acts on line C-X-M and causes the boat to drift toward the towpath—that is, point C (the entire boat) moves on the line C-X-M, the centerline of the boat remaining parallel to the centerline of the towpath all the while. However, if the center of mass is not on the line of the rope, as it is in Figure 13B, then the stern of the boat will skid either toward the towpath, or away from it, because the towrope force is applied to the end of a lever arm (CX).

To visualize how skidding is caused, think of CX as the handle on a pot or skillet, with a pulling string tied to the end of the handle. When the string is pulled, the handle swings on the line of the string; when that happens, the pot must rotate also, just as the boat tends to rotate also until all three points—C, X, and M—are on the same line, as shown in Figure 13AA (or in Figure 13CC). The reader is invited to draw the force diagram required to illustrate the component of force responsible for skidding and, therefore, to estimate the effect of skidding upon the mule.

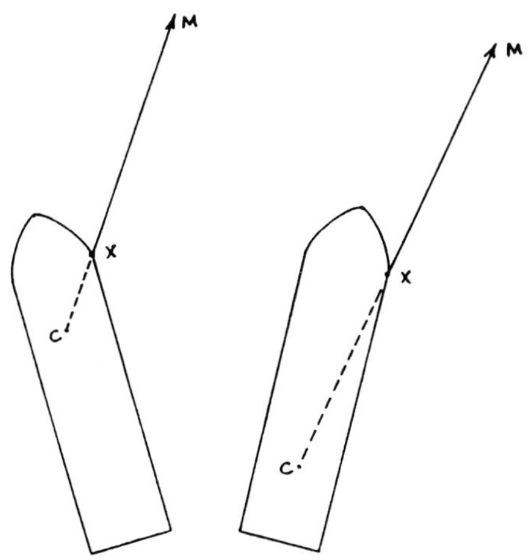

FIG. 13AA FIG. 13CC

We have seen that the towrope applied two forces that caused the boat to rotate about its center of mass: a force exerted sideways upon the front end of the boat, and another force that caused the bow and stern to "skid." A force that causes rotation is called a torque. The tendency to clockwise rotation of the boat, about point C, caused by the sideways force (FS), is counteracted by an equal tendency to counterclockwise rotation of the boat caused by the application of rudder torque. The skidding torque is similarly opposed by application of rudder torque, as required.

For a given towline torque (18 lbs. force, in the example previously used) the effect of that sideways force on the boat is somewhat greater when the rope is attached at point B than when it is attached at point X, because the distance BC is greater than the distance XC—that is, the lever arm or the "cranking arm" is longer in one case, so the application of the 18 lbs. force at the end of that lever transmits a greater rotary moment (force times distance) to the center of mass of the boat. Figure 13 shows the rudder set to apply a counter torque to the stern of the boat, to compensate for the rotation of the stern produced by application of the towline torque to the boat. The rudder torque is applied to the boat at the end of the lever arm RC. The boat pilot holds the tiller (the rudder) in the position shown, and so can feel the amount of rudder force required at any time to keep the boat headed parallel to the towpath. The magnitude of that force depends upon a number of factors that are not the subject of the present inquiry. Note that the rudder had to be kept set in a position such as that shown in Figure 13 even when the boat was gliding parallel to the towpath, because the sideways force (towline torque) was constantly applied to the boat as long as the boat was being pulled—whenever there was tension on the towrope.

The preceding force analysis was not intended to address all of the forces that act on a boat moving through a canal, but only those that affect a mule pulling a boat in a canal. Location of the towpath off the centerline of the boat made a rudder an indispensable part of a canal boat, and also accounts for the unusual point of attachment of towrope to boat. A difference between a barge and a canal boat is the presence of a rudder on every canal boat, a rudder needed because the mule could not walk in front of the boat it was pulling.

PUBLISHER'S COMMENT

Bill Dzombak's Vector-Analysis of the effect of the tow-line pull on the old canal boats is nicely done and explains a great deal about the location of the tow-line anchoring point on the boat. However, it is my guess that our Canal Boat Captains of the Nineteenth Century, unfamiliar with the theory of vector analysis, yet aware of the phenomenon, would look for the "easy way out". What I am suggesting is that these old captains were smart enough not to sit at the stern of their boats constantly pulling sideways on their rudders. Instead they would look for point "X" on their particular boat to attach the towline, where the sideward pull would be fully compensated for, and they could relax at the stern with the rudder parallel to the towpath. Thus the boat could almost steer itself, except in cases where there was a curve to navigate or the boat was cut loose to coast into a lock.

Bill Shank

LETTER TO THE EDITOR

I agree with Bill Shank's comments on mule-towed boats. [See William Dzombak, "Mule Power Force Analysis," and the following "Publisher's Comment," in the November 1991 issue (no. 79) of this bulletin.] As per Captain Frank H. Godfrey, *The Godfrey Letters* (Syracuse: Canal Society of New York State, 1973), tow ropes were 225 feet long and 5/8" in diameter. This would have greatly reduced "towline torque" by reducing the angle to the bank. The mule towline (towpath line) was frequently attached to a deck cleat ¼ to ⅓ of the length of the boat back from the bow. This is evident in the view at the top of page 50 in Harry L. Rinker, *The Old Raging Erie . . . There Have Been Several Changes* (Berkeley Heights, N.J.: Canal Captain's Press, 1984), on page 87 in Thomas F. Hahn, *Chesapeake and Ohio Canal Old Picture Album* (Shepherdstown, W.Va.: The American Canal & Transportation Center, 1976), and on page 30 in Thomas F. Hahn, *The C. & O. Boatmen, 1892–1924* (Shepherdstown, W.Va.: The American Canal & Transportation Center, 1980). On both the Morris and the Delaware/Lehigh canals, a towing post about 4′ high, about 8 to 10′ back from the bow, and offset from the centerline of the boat, was commonly used.

Bill McKelvey

DELAWARE—LEHIGH CANAL CORRIDOR

A Delaware & Lehigh Canal Heritage Corridor Commission (DLCHCC) has been appointed and charged with drawing up a plan for the future management of the area as an educational, economic, and recreational resource. Among the commission members is A.C.S. secretary/treasurer Charles W. Derr.

The commission has already produced a number of attractive publications about the canal corridor. Anyone wishing to be placed on its mailing list is invited to write to 10 E. Church St., Room P-208, Bethlehem, Pennsylvania 18018, or phone 215 861 9345.

BOOK REVIEW

[William H. Shank, ed., *The Canals of New York State* (York, Pa.: American Canal and Transportation Center, 1991).

Reviewed by Jeremy G. Frankel

I'm sure many of you have suffered the same problem, of leafing through all the issues of *American Canals* or thumbing through the four volumes of *The Best from American Canals* to find a particular article, photograph, or map for some research you may be engaged in. That job has now been made a lot easier with the publication of Bill Shank's new book, *The Canals of New York State*.

Included are over 40 articles covering the state that led the way with waterway transportation, including several about what is arguably the most famous (and certainly the most profitable) canal ever built in the United States—the Erie Canal. Other waterways included are: the Oswego, the Champlain, the Chemung, Delaware & Hudson, the Gowanus, the Junction Canal, the Chenango, the Mott-Haven, and the Shinnecock Canal.

As many of these articles have appeared before, any critical comment at this time would be inappropriate. For those unfamiliar with the previous publications (mentioned above) from which some are culled, the choice of articles provides an interesting and contrasting mix of historical accounts, contemporary news items, and modern-day waterway trips and tours. My only quibble concerns the photographs, as reproduction the second time around seems to have made them appear more contrasting. For a student new to waterways, this volume provides a useful and easily readable overview of the most popular canal in America as well as of its lateral connections.

Looking at the map on page 2, one can dream about a second volume covering such waterways as the Black Rock Channel at Buffalo, the Black River Canal near Rome, and the Crooked Lake Canal which, according to an earlier *American Canals* article, is one of the steepest canals in America. And what about the Glens Falls Feeder Canal, without which the Champlain Canal could not function, even today?

As Shank states in his introduction, *The Canals of New York State* is testimony to the enthusiasm, interest, and accuracy of the individuals who have researched and written these articles. As a historian and researcher, I welcome this addition to my library.

The Canals of New York State may be purchased from the American Canal and Transportation Center, 809 Rathton Road, York, PA 17403. The price is $6 plus $1.50 per copy for shipping.

THE CONEWAGO CANAL

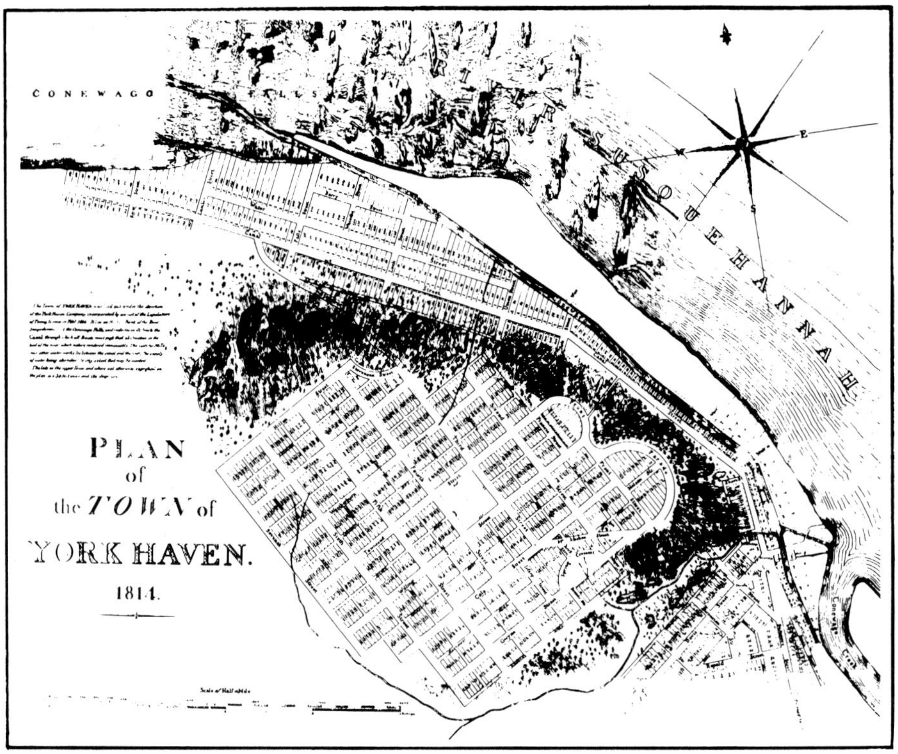

An 1814 map of York Haven, showing the Conewago Canal. The map is the property of, and is reproduced through the courtesy of, The Historical Society of York County.

by George Thomas

[This report is based on an 1802 "Letter to William Penrose, Esq. Chairman of the Committee to Whom was Referred the Petition of Thomas W. Francis for a Toll on the Conewago Canal and Locks on the River Susquehanna." The letter was discovered by the author in the microfilm files of the Stanford University library.]

A letter from Thomas W. Francis to a committee of the Pennsylvania legislature, requesting permission to charge a toll of two dollars per boat, provides an interesting glimpse into the problems that beset one of Pennsylvania's earliest canals, the Conewago Canal. Writing in 1802, Francis, shareholder and treasurer of the Conewago Canal Company, was faced with trying to recover a $105,000 construction cost overrun. His plight will be discussed following the presentation of some background information.

Creating a Canal

In the colonial Pennsylvania of 1771, navigation on the Susquehanna was difficult owing to the presence of obstructions and fast moving water. One particularly treacherous location was Conewago Falls, a stretch of rocks and rapids 14 miles north of Wrights Ferry (Columbia). All goods traveling between Philadelphia and the northern settlements along the Susquehanna had to be loaded and unloaded north of the falls in the town of Middletown. If goods could be diverted further south to the Kings Highway, which stretched from Lancaster to Philadelphia, trade would be improved and shipping costs reduced.

On the 9th of March, 1771, by act of the General Assembly, commissioners were appointed to clear obstructions from the Susquehanna north of Wrights Ferry. Upon examination, however, the job proved too challenging, and the funds appropriated were spent on other improvements. Efforts were renewed after the Revolution, and in 1789 a meeting was held with 26 leading Pennsylvanians, resulting in the appointment of three commissioners "to view and explore the Susquehanna and Juniata rivers, and ascertain the expence, in their opinion, necessary for removing the different obstructions in said rivers."

Their report on January 30th, 1970 called Conewago Falls "the great obstruction and barr to the wealth and population of our western country . . . ." They suggested building a canal 306 perches [rods] long by 33 feet wide, with a drop of 19 feet, rather than attempting to clear obstructions. This canal would be suitable for boats carrying 35 tons with a 20-inch displacement. (They were probably referring to Durham boats, but have overstated their load capacity.) Total cost was estimated at £5,000 for the canal, plus £300 to clear the ripples near the canal entrances.

About a year later, a general appropriation of £5,250 was made by the legislature "For Susquehanna, from Wrights Ferry to the mouth of Swatara, . . ." with no provision for land purchase along the canal right-of-way. The commissioners visualized a canal which followed the edge of the river. On July 3, 1792, Governor Thomas Mifflin contracted with 17 individuals, who formed "The Conewago Canal Company."

If this canal initiative is put into perspective with other events that were occurring in the state, it appears that the granting of this contract was part of a larger plan to improve transportation. Three months earlier, the Philadelphia and Lancaster Turnpike was chartered to build the first hard-surface road in the country. Also, a charter was granted in 1792 to construct a navigable waterway between the Schuylkill and Susquehanna rivers. This undertaking was a short-lived predecessor to the Union Canal, which completed the connection in 1828.

Governor Mifflin inspected the finished Conewago Canal on November 22, 1797, in the company of James Brindley, engineer and superintendent of the works, and other canal company representatives. An appointee, charged with assessing whether the contract had been properly executed, described the weather as particularly severe that day, with ice on the canal and snow in the air. He continued with a description of the ceremonies.

Preceding the official tour of the canal, a quick trip was made by several rafts and a boat up and down the canal to break up the ice. The first locks raised the boat nine feet in eight minutes and subsequently the second lock was passed in two-and-one-half minutes. The difference in time was attributed to unbroken ice in the first lock. The boat was poled through the canal by five people in 17 minutes, three minutes to pass the regulating lock, and five more minutes to make the river entrance above the falls. The official party descended back to the double locks, where the governor landed to a gathering of 500 people. No less than three times, the governor was saluted by the canal workmen, who bored "their artillery out of the hardest rocks" and fired salvos as he passed on the canal. [Since the canal workmen did not have a cannon or other means of making a loud noise to commemorate the visit by the governor, they bored holes in the rocks above the canal, loaded them with powder, and detonated them.]

The following physical description of the canal is included: It is just short of one mile long; It is 40 feet wide at the bottom; It is four feet deep when filled to within 20 inches of the top; The locks are constructed mostly of brick and stone; Each of the lower double locks is 80 feet long by 12 feet wide.

The appointee mentions the cost overrun in his report and advises the governor to consider the canal completed as contracted for £5,000.

The $100,000 Misunderstanding

The plight of Mr. Francis and his associates may now be considered. The financial problems of the Conewago Canal began with the original survey in 1789 by commissioners who had no experience in canal construction. Their £5,000 proposal was to dig a sluice canal around the falls and let the water run through the channel unchecked. The only technical research they did was to ask local boatmen, who assured them that a 19-foot drop would present no problems. When work began on the canal, this approach was considered impractical because the water velocity would have undermined the canal wall and made navigation nearly impossible. The maximum practical descent according to authorities of the day was 500:1 in the length to the drop, or 10.1 feet in 306 perches. The Conewago Falls descent was nearly double that limit. Applying hydraulic principles as recommended by Mr. Francis (although he never did the calculation) suggests the water velocity would be 24 miles per hour. Actual measurements by the canal engineer, Mr. Brindley, show a speed of 12-15 miles per hour, based on the timed descent of a raft at five minutes and of large logs at four-and-one-half minutes.

Considering the reverse case of taking a boat against the current developed in the proposed sluice canal, Mr. Francis proposed using the prin-

Site of double locks looking southeast; this is currently a drainage ditch.

ciple of the inclined plane to calculate how many 180-pound men would be needed. The inclined plane does not seem to be the correct analysis for the situation, but again he offers two analogies based on actual experience:

To bring an empty boat up the Schuylkill Falls, which is three-and-one-half feet in 30 perches (495 feet), requires an hour's hard work by 12 to 15 men. The cost was $2-3 in drink, with no mention of wages.

The few times a boat was brought up the Conewago, 30-40 men were required for most of a day and cost £5-6. The process required the men to jump from rock to rock, pulling the boat. The actions must have been humorous to the local people, since the men were referred to as "blackbirds." Apparently, horses could not be used because of the rocks and uneven shoreline.

With this justification, the canal company rejected the sluice canal concept, as well as one running farther inland which would have required extra digging. The only acceptable alternative was to build a level canal with the total drop taken out with locks. The company chose to build on the west side of the river because it had more water flow and fewer rocks. They agreed that "digging a canal on either side will be a great and expensive work, probably far exceeding the sum allowed by the public . . . ."

The final accounting of the canal expenses follows:

Ground purchase and cutting of canal	$56,726
Locks and regulating gates	45,274
Miscellaneous	17,000
Total cost	$119,000
Amount received from legislature	– 14,000
Cost overrun	$105,000

This concludes Mr. Francis's plea for more money. It appears from the outset that the canal company expected to overspend the allotted sum of £5,250. The fact that construction continued must have been due to the enthusiasm and/or significant borrowing power of the shareholders. Since this was one of the first canals in the country, investors were probably not wary of the cost overruns that would plague subsequent canal projects. The amount of the overrun can be seen in its true light when it is compared to the salary of the lockkeeper that was set in the same report at $200 per year.

The Outcome

Mr. Francis was only partially successful with his petition. According to the *History of Lancaster County* (1883), the legislature authorized a fee of 50 cents per boat for passing through the canal. The company continued to operate the canal at least up to 1814, building a flour mill and operating a ferry next to the canal. The Conewago Canal became less important when the Schuylkill was connected to the Susquehanna at Middletown in 1828, but the last year of canal operation is unknown. In 1886, the *History of York County* reported a papermill was being constructed from the excellent-quality bricks taken from the canal.

Currently, the site of the canal is occupied by a park next to the Pennsylvania Power and Light Company hydro-electric plant. Part of the original canal right-of-way has been lost to railroad tracks and to the water backed up by the power plant dam, but some of the original bed can still be traced. The location of the two lower locks is a ditch used to direct water from a large drain pipe back into the river.

TOWPATHS TO TUGBOATS BACK IN PRINT

Our readers will be interested to learn that the book TOWPATHS TO TUGBOATS, A HISTORY OF AMERICAN CANAL ENGINEERING, temporarily out of print, is once again available. Published originally by a team of ACS Directors in 1982 it has been revised and republished several times. The current issue is the fourth printing, January 1992. This has been an extremely popular book, covering canals and their principal engineers from the Canal du Midi in France, through the early English and American Canals, up to and including the recently completed Tennessee-Tombigbee Waterway. It can be purchased for $7.00, plus $1.50 for shipping, from the American Canal and Transportation Center, 809 Rathton Road, York, PA 17403.

ERIE CANAL BOARD GAME

By William Dzombak

On winter evenings, when the canal is frozen solid, both young and old game board players will enjoy learning how to captain a freight boat from Albany to Buffalo and back, and make a profit, in spite of the many hazards, regulations, and accidents that a boat captain must know how to deal with.

The game board is a map of the Erie Canal, printed on stiff paper, that is 5x34 inches. The map needs to be glued or taped to a rigid backing (cardboard). The pieces or tokens that move on the game board are two boats (folded paper cutouts), one boat westbound (from Albany) and one eastbound (from Buffalo). Movement of boats is determined by the spin of an arrow mounted on a numbered cardboard circle. Each player starts the trip with a stack of paper money used to pay fines (as for speeding, or running a boat that is overweight). Profit made on a trip is calculated by subtracting expenses paid out (tolls, fines) from income realized by selling (auctioning) the cargo at the end of the trip. The cargo carried on each boat is determined, at the start of the game, by drawing "cargo cards" from a stack of face-down cards; the auction price of each item of cargo is found by spinning the arrow pointer.

The game is sold by Chatham Hill Games, P.O. Box 253, Chatham, NY, 12037. The price is six dollars, which is a lot to pay for a large sheet of paper and a plastic spinner, but students of canals are known to attach unusual value to the subject that interests us all. The game could be resold as a money-maker for a canal organization or historical society.

BOOK REVIEW

Ray Spangenburg and Diane K. Moser *"The Story of America's Canals"* (Facts on File, New York, NY, 1992)

Reviewed by Denver L. Walton

As interest in canals increases over the years, more and more publications become available to improve our knowledge. Barely twenty-five years ago about all we had was "Old Towpaths". Now we have before us *"The Story of America's Canals"*, by Ray Spangenburg and Diane K. Moser. This recent effort to present the history of American waterways traces canals and canal building from the earliest ditches through the Panama Canal and the Tenn-Tom Waterway.

The latest in a series on American transportation, this volume is well illustrated with maps and photographs, many from the files of the American Canal and Transportation Center. In the pages of this book we learn such important facts as why the bridges were low and what canal workers were paid.

The Erie Canal is well portrayed as the successful predecessor of the many major canals that followed. The chapter on the Panama Canal covers all the important phases of construction, false starts included, and doesn't neglect the story of the control of the yellow fever mosquito.

This book is available from the publisher, Facts On File, New York, for $18.95 plus shipping costs.

TOURING THE UNION & SCHUYLKILL WATERWAYS

Restored Allegheny Creek Aqueduct at Gibraltar, below Reading.

By Bill Shank

Forty-one members of the Pennsylvania Canal Society from five states convened May 1-2, 1992 at the Riveredge Motel in Reading, Pennsylvania for their regular Spring Meeting and a tour of the eastern section of the Union Canal and the central section of the Schuylkill Navigation.

One of the high spots of the tour was a visit to the new Howard Hiester Canal Museum, near the Reading Airport, which included a professionally assembled and documented canal collection acquired over many years by Howard Hiester, plus a restored covered bridge, two partially restored locks on the Union Canal, a simulated canal boat and a restored wagon works.

Most of our time was spent on the Schuylkill north and south of Reading. The Schuylkill was primarily a slackwater navigation with a series of dams and locks creating a series of pools deep enough for the many canal boats bringing coal from the coal region near Pottsville to Philadelphia. Here and there short canals connected these navigation pools in the river, each having its own name and identity. Many of the dams on both the canals and the slackwater portions are still capable of holding back water. Therefore many miles of the river are still navigable for small boats and canoes, and as a source of water for municipal use.

Our tour included the restored (and rewatered) Aqueduct across Allegheny Creek; a restored Lockhouse adjacent to doublelocks 36 and 37 at Leesport; Felix's Dam and Lock #42; and the flight of Locks (Numbers 52 and 53) at the Laurel Lock Farm near Pottstown.

At the Friday evening session Tour Director Robert Keintz ran a slide show of points of interest on our Saturday tour. Saturday evening, after a satisfying banquet, PCS President Jim Oliver presided over the meeting and told of future plans for the Society.

Bill Shank, First Secretary of PCS, gave a short talk on the events leading to the formation of the Pennsylvania Canal Society in the summer of 1965. Doris Roberts then presented a slide review of a recent tour of the English Canals in which a number of PCS members had participated.

The Union Canal had the distinction of being the first canal route ever surveyed in America. The approximate route for the Union Canal was first conceived by William Penn. Having laid out Philadelphia in 1682, which became the country's first great seaport, Penn issued proposals for another city in 1690 as follows:

"It is now my purpose to make another settlement upon the river Susquehanagh . . . . . which will not be hard to do by water by benefit of the river Scoulkill, for a branch of that river (Tulpehocken Creek) lies near a branch that runs in the Susquehanagh River (Swatara Creek)."

In 1762 David Rittenhouse, the astronomer, and Dr. William Smith, provost of the University of Pennsylvania, made surveys over Penn's route from Reading to Middletown.

In 1792 two companies were chartered by the State of Pennsylvania to build a navigable waterway between the Schuylkill and Susquehanna, improve the Schuylkill, and build a canal from the Delaware to Norristown. By 1794, the companies had completed 15 miles of work, and had spent $440,000, which exhausted their funds, and the work ground to a stop.

In 1815 by act of the Pennsylvania legislature, the "Schuylkill Navigation Company," was chartered to complete the work on the Schuylkill River. The portion of the Schuylkill which was made navigable extended from Port Carbon to Philadelphia, a distance of 108 miles. Of this route, 62 miles were by canal, and 46 miles by so-called "slack water navigation pools" in the river itself, formed by a series of dams passed in each case by locks. Between Port Carbon and Philadelphia there were 92 lift locks to overcome a 588-foot difference in elevation.

The Schuylkill Navigation Company continued to operate from 1825 until 1870 when it was sold to the Philadelphia and Reading Railroad Company, one of its largest customers. Under the new management the system continued operations until 1931.

In the meantime the Union Canal Company, whose responsibility had now been narrowed to the route between Reading and Middletown, was offered financial aid by the state of Pennsylvania. The Union Canal was pushed through to completion between 1821 and 1828.

The restored canal channel, running across the Allegheny Creek Aqueduct at Gibraltar.

Unfortunately, the designers of the Union Canal had made both the channel and the locks of the canal so narrow that the heavy freight boats and large passenger packet boats, on most of the eastern inland waterways, could not use the canal. Special boats able to carry a load of only 25 to 28 tons were constructed for the Union Canal. The canal proprietors secured permission from the state legislature to widen their channel in 1841 but the enlargement was not actually completed until 1856. The Union Canal was then able to handle boats of from 75 to 80 tons capacity, but the company never recovered from this enormous additional expenditure ($6,000,000.) and it was finally abandoned in 1885.

Felix's Dam on the Schuylkill, still maintaining a deep-water pool above Lock Number 42.

BRITISH I.W.A. VISITS THE U.S.A.

By Roger W. Squires

Part of the IWA Tour group on the deck of the "Josiah White" at the Hugh Moore Park in Easton. From the right, Bill Shank; David Stevenson (behind Bill); Bill Moss, President of the New Jersey Canal Society; Roger Squires, United Kingdom Director for ACS (kneeling); Lance Metz, Hugh Moore Park Historian; Ron Oakley, Chairman of the IWA Tour; and Howard Swope, former Delaware Canal Boatman. (Photo, courtesy Linda House, CSNJ.)

Britain's IWA International became of age when it successfully organised its first International Conference in Buffalo, New York State, in conjunction with the A.C.S. on 20th June 1992. The theme was the 'Multifunctional Use of Canals'. Speakers were drawn from the IWA, various American Canal Societies and the Canadian Canal Society. David Stevenson, the IWA National Chairman, joined Bill Trout for the opening address. A second day, run by the Canadian Canal Society, comprised a guided tour of the Welland Canals and included a trip through one of its great ship locks. In all the event was a great success and stimulated those present to ask for a further conference in another year or so.

Some 28 IWA members from Britain attended the Conference, as it formed part of a three week tour of the Canals of the Eastern United States organised by the IWA International Committee Chairman Ron Oakley. Various American Groups were able to use the occasion of the tour to repay past hospitality from their earlier visits to the UK and, as readers will realise, the bonds of international friendship were firmly established whilst at the same time offering U.S.A. canals activists an ideal opportunity to gain maximum publicity for their local canals.

The I.W.A. tour started in Boston with a visit to an early waterpowered iron works and then a dinner in the Middlesex Canal Engineer's House hosted by the Middlesex Canal Association. The overnight hotel in Lowell offered a pair of locks under restoration adjacent to the hotel grounds and that set the scene for the rest of the second day, which was spent touring the town of Lowell and seeing how water power and canals had enabled it to become the first mill centre of America. Whilst the mills no longer operate commercially, it was a welcome sight to see a floor of 100 looms recreated in a restored mill building as part of the revival of the old town as a State Historic Park.

The group was able to see a second State Park when they visited the Blackstone Canal the following day. In this case five local communities have banded together to stimulate the development of a linear park centered on the old and now derelict navigation. A subsequent visit to a major wooden aqueduct on the Windsor Locks Canal and the sight of the Hudson River offered a vision of what was to follow.

View from the bow of the "Josiah White" as the IWA group was dropped through a lock at the lower end of the restored Lehigh Canal.

The rail buffs of the group had a field day at Kingston (N.Y.) where they had a vintage diesel railcar all to themselves on a local restored line. Canal buffs did not miss out as later in the day they were able to explore one of the few suspension aqueducts ever built. It once formed part of the Delaware & Hudson Canal but has since been converted into a road bridge.

The railway centre at Scranton (PA), with its recently rebuilt roundhouse and extensive displays of historic locos, provided a chance for the group to ride on the Lackawaxen Railroad before heading south to the canal town at Easton, Pa. Here the group visited the Hugh Moore Park Canal Centre to see a slide presentation on the PA canals by Bill Shank and to learn about the three major local canals before being given a mule drawn boat ride along the Lehigh Canal. The highlight of the trip was the first hand experience of seeing the 'drop gate' operate on the restored lock.

The New Jersey Canal Society provided the group with a full day exploring the remains of their key canals. Sadly few relics of the grand inclines of the Morris Canal are preserved today but we heard of plans to restore the incline at Waterloo Village. One of the highlights was a BBQ at Jim Lee's where the turbine chamber of the old incline has been excavated and partially restored. The group then went on to see the Delaware & Raritan Canal, with its dropped bridges, and the Feeder at Prallsville, before returning to Easton along the

(Continued on Page 34)

BRITISH I.W.A. VISITS THE USA

The new Visitors Center at the famous Horse Shoe Curve, near Altoona, Pa. The Inclined Plane shown at the left and the walkway to the right take visitors to the upper level where freight and passenger trains make a frequent appearance.

(Continued from Page 33)

remains of the Delaware Canal. The British visitors much appreciated the enthusiasm and knowledge of the local guides for this section of their tour.

It was next the turn of the Chesapeake & Ohio Canal Association members to entertain the group in Washington. My, what a spread they provided. It was a great pity the local replica canal trip boats at Georgetown and Great Falls were not working during that time of the stay. However, that did not detract from the pleasure of seeing the restored locks on the C & O in Washington and at Great Falls. We then went on to see the grandure of the now overgrown early locks on the Potomack Canal at Little Falls. Later in the day we also saw the old lock tenders cottage (now a public convenience) in the center of Washington, (the old canal line here being a 4-lane road.)

The next day our route followed the line of the C & O Canal out to Harpers Ferry (W.VA.) then inland to the site of the Allegheny Portage Railroad in PA, which once transported canal boats over the watershed to the Ohio Basin. All admired the splendid new Visitor Centre. What a pity it was so far from the incline. Railway 'buffs' then had the treat of seeing two trains overtaking on the famous Horseshoe Curve before we headed to Johnstown, PA. The town has two claims to fame. One is the Johnstown Flood and the other is the steepest car carrying incline in the world. The group saw both but could not find any remains of the canal basin around which the town originally was founded. However, this seemed the case almost everywhere in the U.S.A. Once something had been superceeded it is 'trashed'.

Looking down from the Horseshoe Curve at the new Visitor's Center. The Inclined Plane can be seen at the right, carrying visitors between the upper and lower levels. The cars are completely automatic and use a single track, with a turn-out at the half-way point.

David Stevenson (left) National Chairman of the Inland Waterways Association, London, presents a special British Canal "Miniature" painting to Bill Shank on the deck of the "Josiah White" at Easton, Pa., June 13th.

The group then went on to Pittsburgh and at last saw real barge tows and the effective use of inland navigation. Although Pittsburgh is 1500 miles from the Gulf, its wharf areas were receiving regular loads of coal, iron ore and other products. We took a trip downstream to see the huge modern locks that make all this possible and were enthralled to see a push tow with eleven fully laden barges each carrying 2000 tons emerge from the lock. The trip back up the Ohio river presented the vista of Pittsburgh at the junction of the Monongahela and Allegheny Rivers and will provide lasting memories. Quick side trips up the town's two remaining passenger inclines and a ride on the new transit system provided a fitting end to an action packed day.

From Pittsburgh the group headed north to Buffalo for the International Conference and Welland Canals Tour before boarding 'Emita 2' for a five day cruise through the Erie Canal (New York State Barge Canal). This remarkable 365 mile waterway connects Lake Erie to the Hudson River through a series of man made cuts, river navigations and lakes. Sadly today it is hardly used. We were told of the odd oil barge but saw none. What was even worse was the distinct lack of pleasure craft, except on the last few miles at the Hudson River end. Two years ago the Mid Lakes Navigation Company introduced six self-drive-hire boats on the system. These were the first of their kind. We saw them in use; let us hope many more will follow.

The 'Emita 2' is a day boat only, so nights were spent at convenient canalside hotels. Even so all the meals were taken on board during the trip and they were first class. The key features of the first section were the rock cuttings and staircase of two huge locks at Lockport. A rest day was taken midway at Syracuse where visits to the Weighlock Museum and the old Erie line were arranged. This gave everyone a chance to explore a series of three dry docks being re-excavated as well as to see some remains of locks and aqueducts on the old Erie Canal line.

Refreshed by these side trips, the group returned to the 'Emita 2' for the second leg of the tran-

Dr. Roger Squires (left) ACS Director in the United Kingdom, shakes hands with Charlie Derr, ACS Secretary and Treasurer. In the background is the "Josiah White".

The completely new Allegheny Portage Railroad Visitor's Center at Cresson, Pa. Many canal artifacts are housed in this building. A short, down-hill walk takes visitors to the Lemon House and re-built Plane Number Six.

Looking down the rebuilt Plane Number Six on the Allegheny Portage Railroad at Cresson. Strap-rails were mounted on heavy wooden timbers with wooden ties between the up-bound and down-bound tracks. In the far distance can be seen the famous Skew-Arch Bridge.

sit to Troy, N.Y. The size of the lake through which the route passed next was quite daunting. Especially when the Skipper told tales of the way the weather can 'cut up rough'. Still we all survived and many had sunburn to prove it. The route went on past the canal town of Rome, with its restored canalside village, before joining the Mohawk River Valley at Herkimer. Here a local 'diamond' seller joined us for a while.

The transit down the Mohawk Section is quite unique in that the navigation dams are all demountable and can be raised above the flood level to allow the winter ice to move away unhindered. The former carpet mill town of Amsterdam provided the second overnight stop before the final leg to the Hudson Valley. But first we had to transit the famous Waterford Flight of five locks that lowered the canal down the valley side. At the top of the flight we were greeted by Kilted Pipers and all the town dignitaries. Press photos, interviews and then on down the locks. What splendid end to a fantastic cruise. Our berth that night was on the Hudson River at Troy which was reached by passing through the Federal Lock. It was nice to see one of the hire boats moored alongside with people enjoying themselves on this under-used waterway.

From Troy we headed back on the Turnpike to Boston to see the restored Market area and the Harbour. Also the new locks on the Charles River and the trip boats by the Science Museum. The new and old transit systems were explored as was the local railway to Salem. Then all too soon it was back to the airport and across the Atlantic.

What did the IWA visitors learn? Firstly, the small bands of local enthusiasts in the USA all do their own thing. The distances prevent regular easy contact and sharing of ideas. Secondly, much of the early canal network in the USA has been 'trashed'. Roads now take their route. Historic marker Boards often are all that remain. Thirdly, even where there is potential to restore viable lengths for powered boats, the option is not followed through. There is a real fear that powered craft will wash away the banks. The IWA International Conference went a long way toward opening up ideas about restoration, but generally the locals felt it was not viable. Perhaps they are right? It seemed to the outsider that general public attitudes and perceptions in the USA do not relate easily to the early canals. For the most part the odd relics remain as historic monuments each commemorating the canal of which they were once part. One can only hope that the Erie Canal (New York State Barge Canal) will not go the same way.

At the International Meeting in Buffalo, June 20th, 1992: Bill Trout (seated), President of the American Canal Society and David Stevenson, National Chairman of the Inland Waterways Association, United Kingdom. (Photo, courtesy Roger Squires.)

Panelists at the International Meeting in Buffalo, June 20th, 1992. Left to right: John Jackson, David Ross, Roger Squires, David Stevenson, Terry Woods, Bev Morant and Frederica Kleist. (Photo, courtesy Roger Squires.)

GLOSSARY OF CANAL TERMS

Here we have a problem. Do we call this a "flight of locks" or a "lock staircase"? Do we call them "double locks" or "parallel locks"? Impressive, in any case!

A few months ago, A.C.S. President Trout appointed an ad hoc committee to develop a canal glossary. The committee chairman is John W. Droege. The work has now proceeded to the point where the committee is ready to solicit assistance from the A.C.S. membership at large. We are therefore reproducing here, in slightly modified form, a questionnaire which was recently circulated to the committee members. Responses are invited, and should be sent to John W. Droege, Rt. 1, Box 31, Brownstown IN 47220.

Before they tackle the questionnaire, readers are asked to note the following guidelines:

1. The principal focus of the glossary will be on terms which were used during the days of operating canals in the United States, mainly the 19th century. Terms in current use, and English terms in use outside the United States, will be included but are of secondary interest.

2. Because the chairman is already in danger of being swamped with paper, he asks that readers limit themselves to suggesting unusual terms and definitions. He indicates that he already knows how to distinguish a bow from a stern and a lock from a boat.

Following the questionnaire, we present a discussion by committee member William Dzombak of the issue raised in question 7. With the author's customary thorough scholarship, this provides an illustration of how complex these apparently simple terminological questions may be. It may also be taken as illustrative of responses which the committee chairman, given his present paperwork crisis, might consider too verbose. The following alternative model is suggested for reader responses:

Culvert: carries water (possibly a canal) under something (possibly a canal).

Aqueduct: carries a canal over something (possible water).

QUESTIONNAIRE

Please identify your response as "19th Century" or "historical" if possible. If you know of any regional differences in usage, please note. The inclusion of references would be very helpful.

1. Did boatmen refer to two "gates" at either end of the lock, or to one gate with two "leaves"?
2. Were the terms "hold in" and "hold out" (to indicate the towpath or opposite side of the canal) in common use in the U.S.? If not, how did one say to take the near side or far side?
3. What is the difference between "capstan" and "windlass"? Were either or both in common use?
4. Do you know the terms "weir boards" and "weir slots"?
5 "Flash [flush?] lock" — could the gate be opened before the the levels were equalized? Is this the same as "staunch"?
6. What is a "mud sill" with reference to a lock?
7. What is the essential difference between a "culvert" and an "aqueduct"?
a. A culvert has dirt thrown upon the supporting structure, an aqueduct has a wooden or stone channel carrying the water. b. A culvert maintains the usual width of the canal, an aqueduct is just wide enough for one boat. c. A culvert is little, an aqueduct is long. d. A culvert is made of a stone arch, an aqueduct is supported on stone piers. Or e. . . . ? Is there a regional difference here?
8. I know of two kinds of valves: a. butterfly, where a plate rotates through 90 degrees to open or close a passageway for water; b. a plate is withdrawn from before a hole to allow water to pass, or pushed back to cover the hole. I take it the first kind was more common in this country, the second in England. Were there unique names for these two kinds of valve? Is "paddle" or "paddle gate" specific? Does "wicket" refer to one or the other or both? What about "sluice gate"?

"Ground Paddle?"

9. What was the valve called when it controlled water passing through the lock structure rather than through the gate? Was there a name for such a culvert ("ground paddle" in England)?
10. These valves were operated by a mechanism which extended to the top of the gate or lock wall. What was this handle called?
11. In Ohio, at places troubled by excessive silting and inadequate water depth, the state maintained "lighters" to help boats through the bad place. I know what a lighter is in the harbor; what was it in the canal?
12. Prism: was there any fixed relationship among the dimensions? The sides sloped at some fixed angle? The width was 10 x depth?
13. John Lamb sends these terms from Illinois. Were they used elsewhere?
a. "Hydraulic basin": holding basin taking water from the I. & M. (or other?) Canal for use to produce power or for other industrial use, in effect,

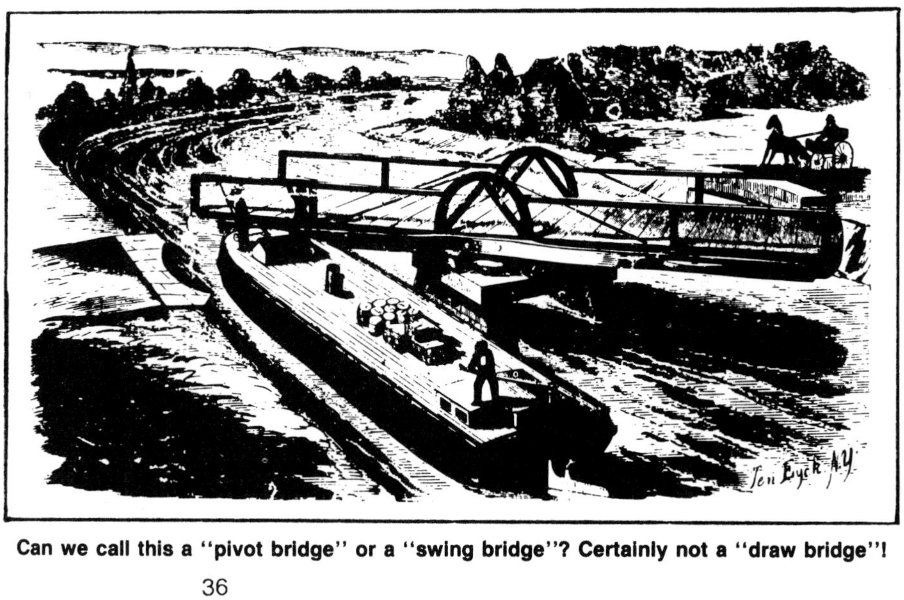

Can we call this a "pivot bridge" or a "swing bridge"? Certainly not a "draw bridge"!

a mill pond. The term "hydraulic canal," I know; "hydraulic basin"?

b. Bear-trap dam: a movable-crest sector dam. One such, formerly located in the Chicago Sanitary & Ship Canal at Lockport, consisted of a 45° sector of a cylinder (perhaps 20 to 25 feet long) of 26-foot radius, pivoted on the downstream side, somewhat in the manner of a Tainter gate. The sector could be rotated to provide an 18-foot variation in the crest. The dam was supported by hydraulic pressure and was operated by controlling this pressure. Its operation was therefore similar to that of the gate in the bear-trap lock formerly used in small streams in the East. Is this the origin of the name? (*Engineering News,* 11/12/1908.)

c. Butterfly dam: formerly located as a safety backup for the above-described bear-trap dam at Lockport. They wanted to be sure not to drain Lake Michigan into Joliet. The structure had two wings, each 80 feet long, supported in the center on a pivot in the center of the canal channel. It operated somewhat like a swing bridge. Each wing had six valves. By opening the valves in one wing, closing the valves in the other, the dam could be caused to swing in one direction or the other, using the current in the canal for motive power. The final movement to or from the fully open position, leaving an 80-foot channel on either side, was accomplished with rack and pinion and electric motor. The dam was normally left in the open position, but was designed to be closed quickly in case of emergency. Was any such thing used elsewhere?

14. What is a "double" lock? Side-by-side? End-to-end? If end-to-end, having a common pair of gates? Please indicate whether you consider your answer to be the 19th century usage.

15. (U.S. usage, please) a. Flight — series of locks separated by ponds; staircase — series of locks with common gates. b. The opposite of a. c. Both terms, interchangeably (series of locks relatively close together). Or d. . . . ?

16. Were these terms used on American 19th century canals: "pound," "winding" or "winding hole," "lock pocket" or "gate pocket," "sweep," "barge" (on towpath canals, "turnaround," "canaller"?

AQUEDUCTS AND CULVERTS

by William Dzombak

A.C.S. member John Droege has begun to compile a glossary of canal terms and has invited assistance from anyone who is able and willing to contribute definitions (and supporting documentation, if possible). To help with that work, the present notice may evoke some discussion and clarification of two terms to be included in the glossary. Those terms are "aqueduct" and "culvert." The task is to identify the essential difference between the two structures.

In common understanding, the difference is that aqueducts convey boats while culverts convey water. An aqueduct is a bridge; a culvert is a tunnel. Confusion may arise if a bridge span happens to resemble a tunnel, as is the case where the trunk of an aqueduct is supported on arches and not on straight (rectangular) piers.

Descriptive definitions. An aqueduct is a bridge supporting a channel, or trough, filled with stationary water. A culvert is a tunnel-like conduit conveying moving water.

Functional definitions. An aqueduct is used to convey boats from place to place. A culvert is used to convey water from place to place.

Elucidation (defense) of the definitions. The most simple form of aqueduct is a single-span bridge; if the bridge happens to span a stream, the span

Water spilling off an Aqueduct into the stream below. Can we call these "waste water weirs" or just "surplus water discharge"???

portion of the aqueduct can be thought of as a culvert, for in such situations the span serves that secondary purpose, also. Most aqueducts were built to cross streams, but other aqueducts were built across dry land, such as ravines, that only occasionally contained flowing water drained from the surrounding hills. Aqueducts have been built to cross dry land where enormous amounts of earth fill would have been needed to level the irregular terrain sufficiently to permit construction of a conventional canal by excavation of the earth to form a channel. Most aqueducts are multispan bridges; structures of that form are never mistaken for culverts.

One or more culverts can be built into an aqueduct, but it is impossible for an aqueduct to be part of a culvert. On some aqueducts that span rivers, each pier supporting the trunk of the aqueduct is pierced by a culvert to enable high flood waters to pass harmlessly through the bridge and not overtop the structure and so damage it.

As applied to canals, the distinction between culvert and aqueduct is simple and clear enough: one is for drainage, the other for navigation. Confusion arises because the same words have been applied in different contexts and there employed interchangeably.

Roman Aqueducts

The Romans, for example, conveyed water from mountain reservoirs over long distances by means of tunnels through mountains and by means of open channels or troughs supported on high piers. The two types of structures used to convey water were referred to as aqueducts, without distinction, if I am not mistaken. The Roman aqueducts ran for great distances before crossing a stream, so it is not necessary to associate stream crossing with the notion of "aqueduct." The Roman usage is also employed today in California, where water is conveyed in huge pipes and in open channels, both called "aqueducts."

Early canal engineers (e.g., Isaac Roberdeau, *Treatise on Canals*—see this bulletin, February 1991, pp. 8-9) referred to a "water bridge," in the sense of what was conveyed (foot bridge, wagon bridge) and not what was spanned. If we use that historic basis for the term, then it is clear that an aqueduct is an elevated and open channel that conveys water over any kind of terrain, wet or dry — that is, a bridge. A culvert, on the other hand, is a closed conduit used to convey water under or through the earth. An aqueduct may, or may not, incidentally serve also as a culvert.

A "large" culvert does not become an aqueduct just by being larger than other culverts. No culvert could possibly serve as an aqueduct, because every aqueduct must be an open channel (to provide clearance for boats), but every culvert is a closed conduit. Some uncertainty regarding this distinction may arise when considering aqueducts made of wood, such as the one at Metamora which is covered with a roof. That covering does not make the structure a culvert, because the roof of the wooden aqueduct is not a necessary part of the structure, but the "roof" of a culvert is indispensable, required to support the earth through which (under which) the culvert passes.

"Large" Culvert Not Aqueduct

A "large" culvert cannot be classified as an aqueduct for another, structural reason, also. Although the span of a culvert may be large enough to warrant classification of the structure as a bridge, that criterion alone cannot justify its designation as the particular type of bridge which is properly called an aqueduct. The difference involves also the nature of the structure found on top of the span. If the span is buried under earthworks that form the towpath, bottom, and berm of a canal, the structure clearly is a culvert; on the other hand, if the span supports a trough bounded by walls and deck made of stone, wood, or iron, the structure clearly is an aqueduct. The distinction here is based on the nature of the materials used to form the channel that lies above the span.

It has also been suggested, in connection with the idea of a size criterion, that a structure which preserves the width of the canal is a culvert while a structure which reduces the canal to the width of one boat is an aqueduct. The distinction rests on the assumption that every aqueduct must be narrower than the canal it serves. Factually, that is not the case, so width of structure cannot be used as the basis for definition.

Definitions that are "obvious" to one person may be objectionable to others, so all canal "experts" are invited to join in the fray. Through discussion may come definitions which will satisfy most, if not all, students of canals.

[Editor's note. A propos of Question 13b., Mary Verhoeff, in The Kentucky River Navigation (Louisville: 1917), pages 37-38, refers to the construction at Beattyville, Kentucky, on the Kentucky River, of "a fixed timber dam with passes on the American bear-trap plan . . ." She states that it was completed in 1886, but proved unsatisfactory, and that in the hope of overcoming its difficulties "a lock was decided upon," construction of which was abandoned before completion in 1891. The index reference is "Bear-trap dam, American.) Citations are the Annual Reports of the Chief of Engineers for 1884 (pp. 1730, 1733, and 1746), 1887 (pp. 1873-76, 1884, and 1888), and 1891 (p. 2449). In The American Canal Guide, Part IV, Trout refers to this as "The BEAR TRAP LOCK," but correctly describes it as "a dam with bear-trap gates (and an unfinished lock)" (p. 16).]

A LONG STRETCH OF QUIET AND PEACE: JUSTICE DOUGLAS AND THE C&O CANAL

Justice Douglas being greeted by interior secretary Douglas McKay, on arrival at Lock #6, March 27th, 1954. Foreground, left to right: Merlo Pusey, writer of the *Post* editorial that started it all; Douglas; Edwward J. Kelly, superintendent of National Capital Parks; McKay; Harry T. Thompson, assistant superintendent of National Capital Parks.

Photo courtesy of National Park Service

By Dave Johnson

"It is a refuge, a place of retreat, a long stretch of quiet and peace at the Capital's back door . . . a place not yet marred by the roar of wheels and the sound of horns." That was how Supreme Court Associate Justice William O. Douglas described the Chesapeake and Ohio Canal in January 1954. The C. & O. Canal has been called "the first national park ever walked into existence." Douglas's description was in a letter he wrote to the *Washington Post,* objecting to that paper's endorsement of a plan to build an automobile parkway on the canal right-of-way. The justice challenged the editorial writer to hike the canal with him. He predicted that if the editor did, he would change his mind and use the newspaper to fight for preservation of the canal. The *Post* accepted this challenge, and the national attention gained by the "Justice Douglas—Washington Post Hiking Trip" of March 1954 ultimately resulted in the creation of the C. & O. Canal National Historical Park.

The C. & O. was born in 1828 during the great national canal-building boom that followed the opening of the Erie Canal. Although it was built and operated by a private corporation, most of the stock was subscribed with local, state, and federal government funds, and the State of Maryland soon became the majority stockholder. The company planned to go up the Potomac valley to Cumberland, Maryland, then cross the Appalachian Mountains into the Ohio River basin. Plagued by construction difficulties, labor shortages, cost growth, and legal obstacles, the canal did not reach Cumberland until 1850. Meanwhile, the Baltimore and Ohio Railroad had reached Cumberland eight years previously, and was now within two years of reaching the Ohio River. The canal company, its construction capital exhausted and in debt, suspended plans for westward expansion and settled down to transporting coal from western Maryland mines to tidewater. Following the Civil War, it experienced a brief period of prosperity. During the 1870s, nearly 600 boats delivered almost a million tons a year to the coal docks in Georgetown and Alexandria.

Financial crisis and devastating floods brought an end to the canal's "Golden Age." The flood of 1889 spelled bankruptcy for the company, and with the state unwilling to rescue it, the canal faced abandonment. Instead, it fell under the control of its old rival. The B. & O. Railroad underwrote its reconstruction and subsidized its operation as the most expedient means to keep the right-of-way from a competing railroad. Under B. & O. ownership, the canal continued to operate until flood damage in 1924 resulted in suspension of commerce.

During the depression winter of 1937–38, the B. & O. used its title to the canal as collateral for loans from the Reconstruction Finance Corporation. Then in April 1938, it offered to sell the entire canal to the government for $2 million, the proceeds to be applied to the R.F.C. debt. The sale was completed in September 1938. The Civilian Conservation Corps then went to work on the restoration of the first 22 miles of the canal, between Georgetown and Seneca, creating a public park to be administered as part of the National Capital Parks system. The reopened canal immediately became a favorite recreational resource, and mule-drawn barge rides were offered from Georgetown in 1941. World War II and the disestablishment of the C.C.C. ended restoration work in 1942. Following the war, new plans for development focused on the possible construction of a parkway along the entire route of the canal.

The Douglas challenge

On January 3, 1954, an editorial in the *Washington Post* urged the government to proceed with the parkway project. "By utilizing the old canal—no longer either a commercial or a scenic asset—it is estimated that the parkway could be built for $100,000 a mile." In response, Justice Douglas stated his opposition "to making a highway out of this sanctuary." He went on to enumerate the canal's "fascinating and picturesque" qualities, and threw down his challenge: "I wish the man who wrote your editorial . . . would take time off and come with me. We would go with packs on our backs and walk the 185 miles to Cumberland. I feel that if your editor did, he would return a new man and use the power of your great editorial page to help keep this sanctuary untouched."

Two days later, an editorial headed "We Accept" replied. "We are pleased to accept Justice Douglas' invitation to walk the towpath of the old canal . . . . We are sufficiently enthusiastic about it to wear some blisters on our feet, but we do not believe that this backyard wilderness so near to Washington should be kept closed to those who cannot hike for twenty miles a day."

Douglas's name quickly attracted leaders of major conservation organizations who wanted to join the expedition. Eventually, some three dozen were invited to participate. Logistic support was provided by the Potomac Appalachian Trail Club, and arrangements for overnight lodging were made with sportsmen's clubs along the route. The hikers' gear was transported by truck, so they would not be burdened by heavy packs. The National Park Service assigned the chief naturalist of National Capital Parks and park policeman to accompany the group. Meals were prepared by P.A.T.C. volunteers or by the host clubs. The overall coordinator of the hike was Jack Durham of the Wilderness Society.

The *Post* was represented by the author of the editorial, associate editor Merlo J. Pusey, who at 52 was only three years younger than Douglas, and Robert Estabrook, 35, editor of the editorial page. The regular outdoor reporters of the *Post* and the *Evening Star* sent back daily reports to Washington, and *Time* and *Life* magazines and the wire services provided national coverage.

Eight days that shook the capital

The party left Washington on Friday, March 19th, on a special car attached to the B. & O.'s Chicago Express, and arrived in Cumberland for a kick-off banquet hosted by Maryland Senator J. Glenn Beall. (Beall was a strong advocate of the parkway, as were most of the people in Cumberland.) The next morning, under threatening skies, the hike started at Lock 72, bypassing the first 10 miles of the canal. Following picture taking and interviews, 37 hikers started down down the towpath at about 8:30, and it is reported that 34 reached the Cardinal Club near Town Creek, below Oldtown, the scheduled stopping point for the night.

The second day, Sunday, took the group another 21 miles, to below Little Orleans. The day started with heavy flurries and a stiff wind, and for the first two hours they walked in snow. "Scat" Eaton, the son of a former canal boatman, met them at Paw Paw with a lantern to lead them through the tunnel. They found that much of the original wooden towpath beyond the downstream portal had rotted away. In one place, only a single narrow plank supported by iron pins in the rock wall allowed them to avoid wading in freezing water two feet deep. Farther down, the Cumberland Outdoor Club invited the party in for ham sandwiches and beer. This was greatly appreciated, but since it was only 9:30 in the morning, the break disrupted the day's pace. Two dozen of the original starters checked into the famous Woodmont Rod and Gun Club late in the afternoon for a buffet dinner.

On Monday, the *Post* carried an editorial in which Estabrook and Pusey stated, "we are not at this point in the journey trying to draw final conclusions. We are, however, impressed by the general demand for better use of this natural playground and the obvious eagerness of many people to get away from the pressure, tensions and exactions of city life. Out here the first signs of spring seem far more important than the antics of self-inflated wild men or what Congress does with the tax bill . . . . At this point we are torn between a feeling of appreciation to Justice Douglas for luring us into this venture and irritation over the increasingly pathetic condition of our feet."

With weekend hikers back at their desks, a smaller group pushed on for 22 miles on Monday, reaching Fort Frederick. Much of the towpath in this area was heavily overgrown, in some places so much so that they had to walk in adjacent fields, and considerable old flood damage was found. That night the cooks prepared the evening meal over an open fire, and, in 20-degree temperature, they camped out for the first time. Only 15 remained who had walked all the way. Nine others were still with the group, but had hitched rides on the truck for parts of the trip. (This latter group included Estabrook and Pusey.)

The Fourth Day

The fourth day took the group to Falling Waters, below Williamsport, where Lee's army had retreated across the Potomac after Gettysburg. (One of the hikers renamed the location "Falling Arches.") During the day, they met 81-year-old Ezra Bangoff, who told Douglas that he had worked on the canal 66 years before, for 50¢ a day, first as a mule driver and then as a steersman. A school band played as the sore-footed troop passed through Williamsport. They spent the night indoors at the Potomac Fish and Game Club.

Ten hikers still had perfect records at the end of day five, with another 12 "duffers" still in the party, when they reached the Conococheague Sportsmen's Club, a couple of miles below Shepherdstown. Justice Douglas continued to maintain his rapid pace and was among the first three to arrive, after a 24-mile walk in eight and one-half hours. The leaders were reported to be Constant Southworth, 59, and George Frederick Miller, who at 73 was the oldest member of the party. (According to George Kennedy of the *Evening Star,* Douglas would have been ahead, but he had to stop to sign autographs at every road that reached the towpath.)

The next morning, in the rain, they started on what turned out to be the toughest leg of the hike. The 24-mile segment to Point of Rocks included a towpath break near Harpers Ferry, where the rising river inundated the canal. To pass, they had to climb 150 feet up the steep side of the mountain and clamber along the exposed rock faces. Douglas, who in recent years had hiked 240 miles in the Himalayas, went up the mountainside using grape vines to ease his way for several hundred yards to the point where he could return to the towpath. Twenty-five hikers checked into Camp Kanawha for a steak dinner. Nine were still bona fide every-step-of-the-way hikers.

Justice Douglas in the lead on the towpath. (Photo courtesy of National Park Service)

On Friday, the temperature climbed into the 70s, and, reinforced by weekend recruits, they added another 24 miles to arrive at Seneca in nine hours. They dined that night on buffalo steaks at the Izaac Walton League clubhouse near Poolesville.

That final night, the hikers formed an association to keep alive the friendships created by the experience. To continue the effort to influence development of the canal, they appointed a committee to summarize their conclusions and recommendations. Douglas was named chairman of the committee, which included Estabrook; Olaus Murie, president of the Wilderness Society and a noted biologist; Harvey Broome, attorney and vice president of the Wilderness Society; Bernard Frank, assistant chief of watershed management research for the U.S. Forest Service and chairman of the executive committee of the Wilderness Society; Howard Zahniser, Wilderness Society executive secretary and editor of its magazine, *The Living Wilderness;* Sigurd F. Olsen, president of the National Parks Association; Anthony Wayne Smith, assistant general counsel of the C.I.O. and member of the executive committee of the National Parks Association; Irston Barnes, president of the Audubon Society of D.C.; George Blackburn, president of the Potomac Appalachian Trail Club; Bill Davies of the U.S. Geological Survey; and Lewis Shollenberger of C.B.S. Radio news.

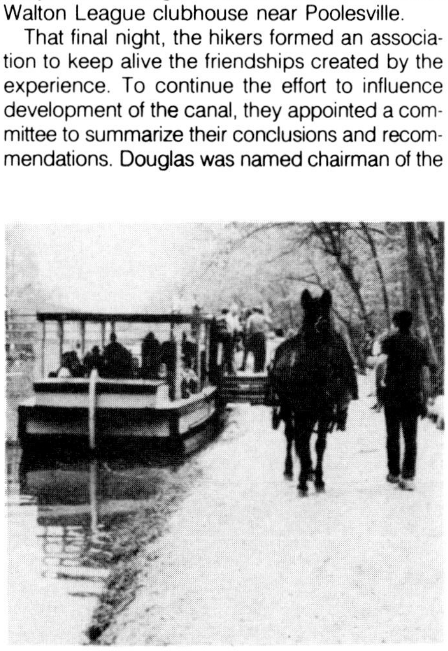

A canal boat loads at Great Falls, circa 1970. (Photo by Bill Shank)

The next day, Saturday, March 27th, was the triumphant conclusion of the hike. Walking the 18 miles from Seneca, picking up followers along the way, they reached Lock 5 at Brookmont in mid-afternoon, to be greeted by Interior Secretary Douglas McKay. "Justice Douglas, I presume," smiled the secretary. "You've got some wonderful country up there," replied Douglas. "I'm going to write you a little letter." The party then boarded the Canal Clipper to cruise the rest of the way to Georgetown behind the mules Dick and Dinah. Traffic on Canal Road was jammed as the boat passed and a large delegation of citizens was on hand when it arrived at Lock 4. Several canoes followed the barge into Georgetown. One carried signs with the messages "Scratch the Parkway at the Post" and "May Justice Prevail." (Another sign, tacked to a tree, said "Jackasses have traveled this path before.") Douglas told interviewers, "I felt that anybody who'd walk this towpath and see its beauties in the raw would never want to turn it into a highway." He added that he was in favor of access roads and outdoor shelters along the canal to entice more people into the wilderness.

Nine members of the party had hiked all of the way, and became known in modern canal lore as the "nine immortals." These were Douglas, Murie, Broome, Southworth, Miller, Grant Conway, John Pearmain, Colin Ritter, and Al Farwell.

Reaping the harvest

Several days after the hike ended, the editorial page of the *Post* carried "A Report." "We retain the conviction that the valley ought to be opened up. We belive, however, that a compromise is possible which will preserve large areas . . . and still make possible a parkway along some beautiful parts of the valley . . . . We think it is possible to develop this resource so that it will serve as a boon to hikers as well as to those whose enjoyment of nature must be limited to a leisurely drive in an automobile."

The following month, Justice Douglas wrote to Secretary McKay, outlining the ideas of the committee assembled on the final night of the hike. The committee proposed tha the canal be preserved as a recreational and historical entity within the national park system. As for the parkway, they recommended that it be developed following existing roads where practicable, perhaps at places parallel to, but never on, the canal, and wherever possible on high ground, safe from floods.

Secretary McKay replied, in May, with the significant remark that "I was delighted to find that the suggestions presented by your committee so closely parallel those of this Department in so many particulars. Indeed, it appears that there is complete agreement on the major objectives to be achieved."

(Concluded on Page 40)

DOUGLAS AND THE C&O CANAL

(Continued from Page 39)

In 1955, the Douglas hiking party held a reunion. About thirty of the original hikers camped below Paw Paw at the Cumberland Outdoor Club (scene of the previous year's mid-morning beer break), then hiked to Little Orleans, and wound up the weekend with a banquet in Hagerstown. Justice Douglas announced that he hoped to make the reunion hike an annual event, beginning a tradition that has continued to the present time. The following year, they formally organized the C. & O. Canal Association as an expansion of the committee formed on the 1954 hike. The association sponsored the annual reunions and spearheaded the efforts to bring about enactment of the national historical park bill. During the 1960s and 70s, the annual hikes grew into major media events, attracting hundreds of walkers, including senators, members of congress, and cabinet officers, to follow Douglas along a dozen or so miles of towpath each spring, and generating continued public interest and support for a canal park.

In 1956, Interior proposed creation of a national historical park, with a separate parkway away from the canal between Hancock and Cumberland. Legislation to establish the park was introduced that year by Senator Beall and Representative DeWitt Hyde of Maryland. Additional bills were introduced in every Congress through 1970. Most passed the Senate but died in committee in the House. Finally, a bill introduced by Maryland's Senator Charles Mathias and Congressman Gilbert Gude was passed by Congress in December 1970.

The 1971 hike, from Spring Gap to Cumberland, was a victory celebration, because President Nixon had signed the national park bill on January 8th. More than 600 people attended. Douglas, then 72 and with his heart regulated by a pacemaker, walked only about two miles. It was the first time he had not finished the scheduled walk.

For the 20th anniversary in 1974, the association held its first full-distance reenactment of the

Dick and Dinah, followed by a crowd of well-wishers, pulling the *Canal Clipper* into Georgetown with the hikers on board. (Photo courtesy of National Park Service)

Justice Douglas relaxes on the boat. Estabrook and Pusey of the *Washington Post* occupy the seat ahead of him. (Photo courtesy of National Park Service)

original hike. A complete hike has been held every five years since then. One major change has been the allocation of two weeks in lieu of the eight-day marathon that Douglas led in 1954. The traditional one-day hikes have continued annually, and in the 1980s a fall "heritage hike" was added to the association's program. In addition, the association sponsors many informal walks and other events throughout the year, and has an active Volunteers-in-Park group. Its Level Walker program assigns members to sections of the towpath which they serve as trail stewards. The association, with about 1,000 members, continues to work as an advocacy group to protect and defend the C. & O. Canal.

Justice Douglas continued to attend and lead the annual reunions as long as his health allowed. In 1977, Congress dedicated the park to him, and a small bust was placed next to the canal at Lock 3 in Georgetown. He died in 1980.

In 1971, Justice Douglas wrote, "When perplexed with legal problems, I have for years taken to the old towpath, hiked a spell, sat for awhile on a bluff overlooking the river, and, escaping from people, lost myself in reveries. Hiking or just sitting in the solitude of the woods and river has some magic. Why, I do not know. But exercise and the quiet of cliffs, woods, and river generate powerful subconscious forces, and before I get home the seemingly insoluble problem has been solved."

Today, the C. & O. Canal National Historical Park provides this "long stretch of quiet and peace at the Capital's back door" for millions of people, because of the dedication and perseverance of Justice Douglas and the little band of conservationists that carried the fight for so many years. The C. & O. Canal Association's Douglas Reunion Hike is held each year on the last Saturday of April. The association and the National Park Service are looking forward to hosting the 1992 International Conference on Historic Canals in October, at Harpers Ferry. For additional information on both these events, readers are invited to write to the association at P.O. Box 366, Glen Echo, Maryland 20812.

BOOK REVIEW

Linn Loomis, *Here and Now—Ohio's Canals: The Background of Ohio's Canal System* (Newcomerstown, Ohio: Linn Loomis, 1991).

Reviewed by Robert H. Mueller, Jr.

A first reaction to the title of this work might be "Oh! another canal book," but first impressions should not be allowed to mislead potential readers into passing over this one. It is worth the time.

The author has expended much time, energy, and resources, to travel around Ohio, converse with many people, and take many pictures, while exploring the canals, and then compiling all these data into book form. The book is organized into literary and pictorial sections, with the former explaining the whys and wherefores of the canals of Ohio. The latter section illustrates many remains of these canals.

A seasoned "canawler" can visualize the descriptions in the text, and also recognze many of the sites in the photographs. Other readers should find the text and the pictures interesting, until they tire of trying to follow the "hopscotching" in both sections, which are treated independently, with no continuity between them.

The hopscotching effect is created by skipping around to the various sites, to illustrate examples of canal and lock construction. The author might have done better to focus on either the Ohio and Erie or the Miami and Erie for the examples, since both were built in the same time period and to the same specifications. This would also apply to the branch canals, even though they were mostly built by private interests.

The text describes many features of not only canals, but also special parts of the locks and the channel sections, that can enrapture the reader with its details. What is confusing is the frequency with which references for a particular detail skip around to widely separated sites of the state-wide canal system.

The pictorial section consists of snapshots of the remains of the Ohio canal system, taken by the author between August 1977 and February 1991. What must be realized by the viewer is that what is shown in the pictures may be entirely different today, if it still exists at all, particularly in the older pictures. Too many of the photographs appear to be out of focus, which could be partly attributed to less-than-ideal conditions when the pictures were taken. The captions explain the subject matter, but here again the hopscotching effect is noticeable.

Sketches or, at the least, rough drawings, would have been an excellent supplement to both the text and the pictures. It is difficult enough to describe anything in words without overwhelming the reader. It is just as difficult to have all factors, many of them variable and uncontrolled, to coöperate when a picture is being taken. A sketch or drawing can go a long way to clarify or explain a point.

The "Bibliography of Canal Reference Materials," near the end of the book, gives the reader a wealth of additional material about not only Ohio's canals but also other canals as well. Local libraries should have most of this material for reference, if not for check-out. Following the bibliography are several pages of additional photographs, unsystematically arranged, perhaps included as fillers.

The book deserves a place in the library of anyone interested in canals. To find a particular picture, or part of the text, would be difficult unless the owner had indexed his copy. This reviewer found himself looking through and rereading the text on more than one occasion.

INTERNATIONAL CANAL CONFERENCE AT HARPERS FERRY

The C and O Canal Aqueduct across the Conococheague Creek at Wiliamsport, Maryland. The upstream wall collapsed shortly before the canal was abandoned, and it was never rebuilt.

Eight-six canal buffs assembled at Harpers Ferry, West Virginia for the International Conference on Historic Canals, October 18-22, 1992. The five-day affair, held at the Cliffside Inn, was co-sponsored by the C. & O. Nat'l Historic Park (Sharpsburg, Md.) and the C. & O. Canal Association, (Glen Echo, Md.) Support was also provided by the American Canal Society and the Cuyahoga Valley Association. There was an "Early-Bird" visit to the town of Harpers Ferry and nearby C. & O. Canal Locks 33 and 34, on Sunday, followed by a reception with canal music by Edie Roudebush.

Bill Trout was "Em-Cee" at the Cumberland Banquet Tuesday evening.

Charlie Derr demonstrates the procedure for putting new shoes on one of the mules at Great Falls.

Monday morning sessions included history and development of the C. & O. Canal Park and problems involved in its maintenance. In the afternoon we visited the Paw Paw Tunnel, and several other points of interest at the west end of the C. & O. Canal. Dinner was provided at the Holiday Inn in Cumberland, with Bill Trout, President of the American Canal Society, as "Em-Cee" and Chris Brown of the National Park Service as speaker.

Tuesday morning the participants assembled at the Cliffside to hear about canal park developments along the C. & O.; the Virginia canals; and the Illinois and Michigan National Heritage Corridor. The afternoon field trip included visits to the ruins of the Canal Aqueduct and Lock

(Concluded on Page 43)

BOOK REVIEW

Terry K. Woods, *Twenty Five Miles to Nowhere*, 2nd ed. (Coshocton, Ohio: Roscoe Village Foundation, 1991).

Reviewed by Lorle Porter

Jack Giech writes that Terry Woods knows more about the Walhonding Canal— "That unfortunately truncated venture"—"than anyone else alive today." That knowledge, and Woods' lifelong love affair with canal lore, shows in this revised edition of *Twenty Five Miles to Nowhere*. The slim volume is packed with every conceivable piece of information ferreted out in several decades of studying this "boondock-nourishing feeder canal" of the Ohio-and-Erie system.

Woods is an engineer, and his description of the building of the canal manages to enlighten an amateur without insulting experts. Along with the fascinating nuts and bolts of the engineering of the canal, Woods captures the flavor of the canal-building era, but he is especially intrigued by the impatience of those left out of the boom of the 1830s. Intense lobbying produced Ohio's "first state road" 29 miles from Millersburg to Port Washington in 1833. More elaborate plans to connect the outland included plans for many feeder lines. In 1836, the state Board of Public Works authorized several major projects including the Walhonding and Mohican Canal.

Work on this canal from Roscoe up the Walhonding River was hampered by financial panics, rising costs, and an increasingly wary legislature. Attempts by local merchants partially to fund the canal collapsed with the expansion of the railroads. The Milan Canal, a legislative darling, received funding and by 1839 was shipping wheat out of the Walhonding region on lakegoing schooners. Enthusiasm for the Walhonding Canal — such as it ever was — crumbled.

A major problem with the canal was its legal description: the 1836 authorization called for a canal from Roscoe "upstream as far as the Board of Public Works deems necessary." "Upstream" was a distance of twenty miles to the forks of the Kokosing and Walhonding — and no major town. Millersburg was up one branch forty miles and Mount Vernon up the other forty-one miles. The canal ended in a slackwater pool 1.2 miles from Rochester and 1 mile from Cavallo—both barren villages. Boatmen had to pole across the slackwater to either "port." Woods finds no evidence of even a bridge to connect Knox County customers with the canal.

Completed, the canal did little business. Subject to one of the wildest of Ohio rivers, it was frequently closed for repairs. "The best built canal in the state" was an economic embarrassment. The Toledo, Walhonding Valley and Ohio Railroad, built in 1893 connecting Coshocton and Loudonville, illegally ran up many of the towpaths — to the considerable joy of merchants, who cared little for the canal. Soon after, the $600,000 state investment was appraised at $9,110.40 and sold off in small parcels to farmers, except for eight miles of feeder used to supply water leases in Roscoe Basin.

Historians are used to discussing the glory days of the canals, but the Walhonding did not even enjoy those. Woods raises interesting questions of why. Efficient competition from the Milan Canal? Lack of entrepreneurial spirit in the valley?

Twenty Five Miles to Nowhere can be purchased from the Roscoe Village Foundation, 381 Hill St., Coshocton, Ohio 43812, for $3.95 plus $1.50 shipping and 24¢ tax for Ohio residents.

REOPENING OF CANAL HINGES ON FLOODWALL GATE

This perspective shows part of the proposed route of Richmond's restored canal. It depicts the canal as it heads eastward under the A.C.L. Railroad trestle. The double-decked bridge is the Richmond Metropolitan Authority Expressway (I-95). The proposed floodwall gate would be just beyond the distant bridge piers. (Drawing by Carlton Abbott.)

The Canal Committee of Historic Richmond Foundation is now conducting a campaign among Richmond's private and public sectors to raise the approximate $500,000 cost of a gate for the James River and Kanawha Canal in the floodwall now under construction. Prompt action to provide the construction funds for the gate by mid-fall is critical to the future reopening of the canal through downtown Richmond, a leader of the movement to restore the canal told the city council on July 8th. Reopening the canal is a key part of the current city and Richmond Renaissance proposal for revitalizing Richmond's waterfront, said A. Howe Todd, chairman of the Canal Committee.

The committee strongly supports Richmond Renaissance's concept for incorporating the Richmond canals into its proposal for the waterfront, Todd said. The most critical issue at this time is the difference in the cost of building the gate *now* (while the floodwall is being constructed) rather than *later,* he emphasized. Some estimates indicate that future installation of the gate into the completed floodwall could cost four or five times the cost of building it now.

Todd's call for action came during a presentation of a detailed engineering plan for reopening the canal's eastern end, which will be intersected by the new floodwall. The study deals with the sector from 17th Street to Virginia Street—an area where portions of the canal system are missing or covered over. The report, which emphasizes the exceptional development potential of the proposed canal restoration, was commissioned by the committee with funds from the Richmond Industrial Development Authority.

Todd said the long-awaited cost estimate was received from the Corps of Engineers only on July 8th, and that the half-million-dollar figure was a rough estimate. A firm estimate was expected to be received in August. Scott said that an immediate effort would be made to raise the full amount from the public and/or private sectors. He said it would be unthinkable for the citizens of Richmond to sit by and allow construction of the floodwall to be completed without the gate being included, and stressed the need for prompt action.

The report, prepared by Carlton Abbott and Partners, P.C., adds engineering data and detail to a schematic canal plan completed in March 1988 for the Canal Committee, also with funds from the Industrial Development Authority. The I.D.A.'s charge to assist in the economic development of the city has been the impetus of its continuing financial support of the committee's efforts.

In praising the recent Richmond Renaissance riverfront report, Todd quoted Brenton S. Halsey, chairman of the Renaissance Working Committee, calling the development of the riverfront and canal "an extraordinary opportunity for economic development, unmatched in potential by similar projects in any other major U.S. city."

Reopening the canal and implementing the riverfront plan, Todd said, is the most dramatic and doable economic development solution for downtown Richmond, providing a uniquely appealing and historic urban amenity as an attraction for new commercial activity and tourism. In the portions of the downtown area where the floodwall runs between the river and the canal, the canal will provide a water amenity, he said. He noted that the potential of such a navigable waterway running through an urban center is dramatically demonstrated in the economic boon to San Antonio, Texas, of its developed canal.

Contributions for the canal gate may be sent to Historic Richmond Foundation, 2407 E. Grace St., Richmond VA 23223.

ERIE CANAL IMPROVEMENT BOND

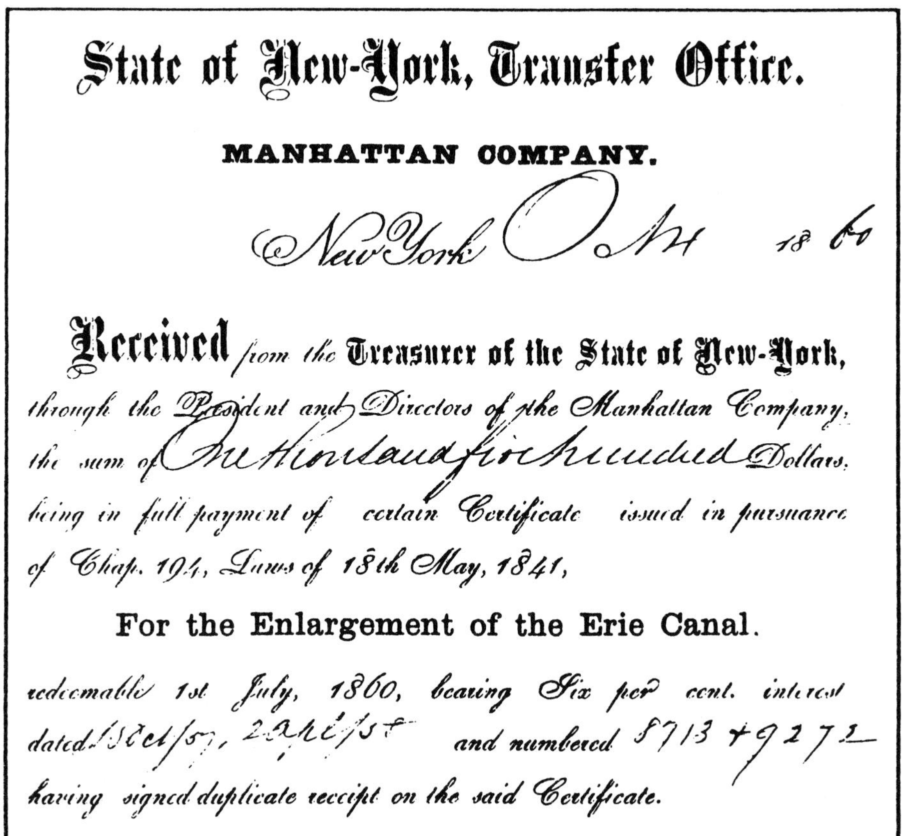

State of New-York, Transfer Office.

MANHATTAN COMPANY.

New York Oct 24 1860

Received from the Treasurer of the State of New-York, through the President and Directors of the Manhattan Company, the sum of One thousand five hundred Dollars, being in full payment of certain Certificate issued in pursuance of Chap. 194, Laws of 18th May, 1841,

For the Enlargement of the Erie Canal.

redeemable 1st July, 1860, bearing Six per cent. interest dated 1 Oct/57, 2 Apr/58 and numbered 8713 + 9272 having signed duplicate receipt on the said Certificate.

A satisfied investor signed this receipt for $1,500 when he redeemed his matured bond for the enlargement of the Erie Canal in October 1860. The facsimile was sent us by Gerard Chapman of Stockbridge, Massachusetts.

INTERNATIONAL CANAL CONFERENCE

The tour group visits Locks 45 and 46 by-passing Dam Number Five on the C and O canal.

The International Canal Meeting attendees ride the canal boat at Great Falls, Maryland, back of mule power.

Members of the Park Service staff field questions on the operation and maintenance of the C and O Canal facilities.

The Monocacy River Aqueduct on the C and O Canal. The masonry is in excellent condition and the Park Service is stabilizing the entire structure to keep it that way.

(Concluded from Page 41)

44 at Williamsport, Maryland, Dam #5 and Lock #45 and #46 (adjacent) as well as "Big Pool" and Fort Frederick, all west of Williamsport. Back at the Cliffside, the evening banquet was followed by a very popular "Show and Tell" program, chaired by John Frye, which lasted till almost midnight.

Wednesday morning the group heard an interesting discussion of the present day activities of the Ohio and Erie Canal Park by Rory Robinson of the National Park Service, followed by a talk titled "Canals in Canada" by John Bonser of the Canadian Park Service. Problems developing along the routes of the Indiana Canals were discussed by Robert Schmidt, President of the Canal Society of Indiana. We ate a box lunch that day at the Monocacy Aqueduct on the C. & O., followed by a tour of canal park facilities at Great Falls, Maryland and a canal boat ride back of mules. Dinner that evening was held at a near-by restaurant, where Tom Brock, Waterways Manager for the Stratford and Grand Union Canals of England, was the speaker.

The meeting concluded Thursday morning with a tour of the Antietam Battlefield, (Sharpsburg, Maryland) led by John Frye. The weather was cool but pleasant for the entire five days, and everything was well organized and co-ordinated by Dave Johnson and his committee — a most worthwhile and enjoyable affair! (W.H.S.)

Lock Number 44 on the C and O Canal, east end of Williamsport, Maryland. This lock is being stabilized and the by-pass flume at the left is being rebuilt.

We visit Lock Number 75 and its rebuilt Lock House at North Branch near the west end of the C and O Canal.

THE ZOAR FEEDER AND SIDECUT

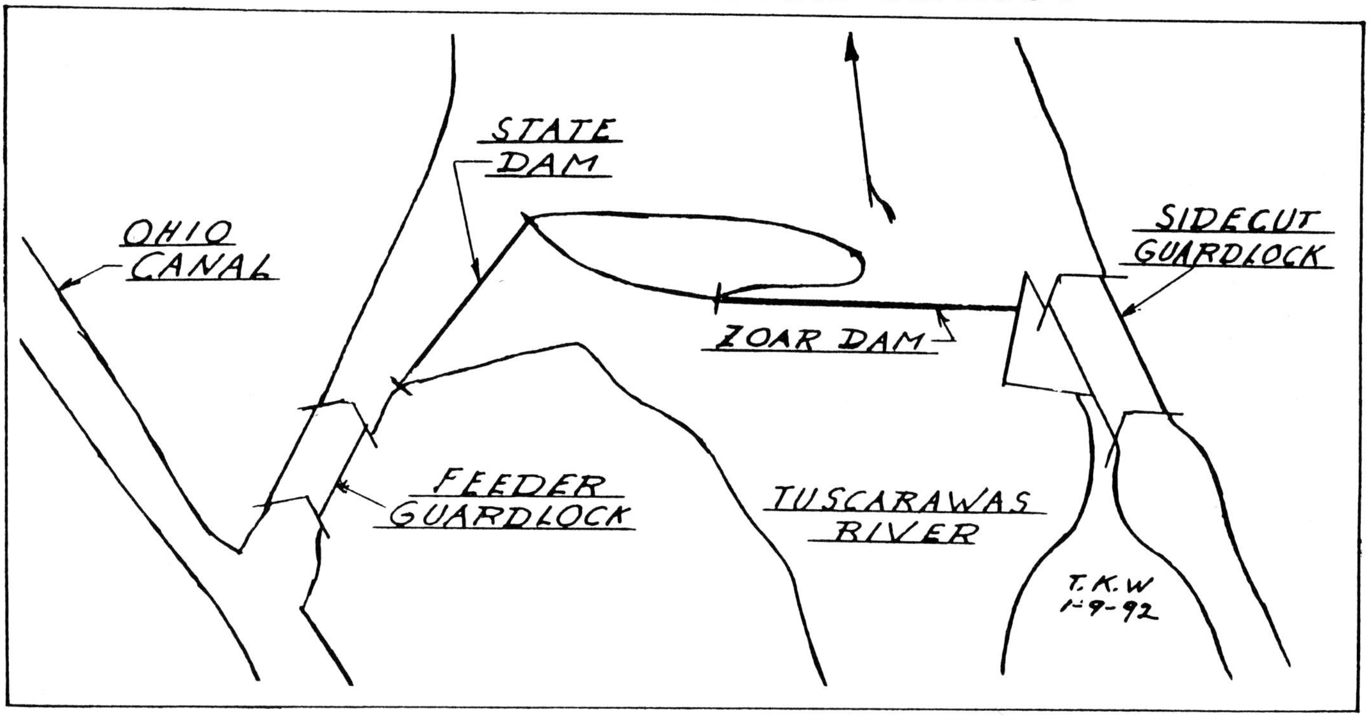

by Terry K. Woods

That seven-mile stretch of the Ohio and Erie Canal that runs from south of Bolivar to near the current crossing of route 800 was constructed for the state by the Separatists. These people came to the wilderness of Ohio from what is now Baden-Württemberg, in Germany, seeking religious freedom. The first contingent arrived on the east bank of the Tuscarawas River, about three miles south of its confluence with Big Sandy Creek, in October of 1817. Other Separatists soon followed, and a settlement was founded on an elevated plain overlooking the river. It was named Zoar after the town where Lot found haven after he fled from the sinful cities of the plain.

By design, the settlers kept apart from the other villages in the region. They worked toward becoming self-sufficient. A dam was constructed across the river in 1821, about a quarter of a mile west of town. A large grist mill was built on the race that was dug from the dam southward, skirting the village on the west. Several other local industries were also begun. Ohio winters are harsh, however, and the mortgage on their 5,500 acres of good Ohio land loomed over the new community. A first payment of $5,000 was due in 1828, a second the year after, and a third, a final payment of $6,500 was to be made in 1830. The Zoarites felt they could accumulate enough money for the 1828 payment. How the remaining two were to be handled was a problem that demanded much prayer.

The Zoarites believed strongly in prayer, but they also believed strongly in doing what had to be done to help prayer along. Beginning in 1819, the society was operated as a commune. All worked as hard as they could, but reaped from the common benefits only the amounts required for their maintenance.

The town of Zoar, looking north-northeast. The sidecut is less than 300 yards due west.

All of the settlers, men and women, were required to work in the fields and the fledgling industries to support themselves and earn the money to pay off the mortgage. Fully two-thirds of Zoar's population were women. To insure that these women would be available to work as required, a "law of celibacy" was issued for all the people of Zoar, beginning in 1822. Then, in 1826, the prayers of the Zoarites were answered. Representatives of the state met in Kendal (Massillon) that year to plan the exact route and issue contracts for extending the Ohio Canal south along the valley of the Tuscarawas. The contractors selected for this work would be paid in cash! The Zoarites bid for this potential godsend—and won.

The entire population of the village turned out during the working seasons of 1827 and 1828. Those who weren't absolutely needed at home to insure an adequate harvest and supply of food for the coming winters, worked on the canal. Many of the canal workers during these years were women. When the lack of wheelbarrows to move the excavated earth became a problem, they carried it to the spoil banks in small wooden tubs balanced on their heads. On occasion, it is reported that they even carried spoil from the excavations in their aprons. In addition to seven miles of canal channel, four stone locks and bypasses, plus several stone culverts, were constructed in this stretch by Zoar's expert craftspersons.

A supply of water was needed for the canal in this section. This required the construction of a feeder dam across the river above the existing Zoarite mill dam, a short stretch of feeder channel, and a gate to regulate the flow of water from the river's slackwater pool into the main canal. These early canal dams, as specified by the state, were hardly worthy of the name, being little more than cut trees, brush, and stone laid in a pile across the width of the river. They did back up sufficient water to feed the canal, but such dams had to be renewed after nearly every flood.

The Society of Separatists received $21,000 from the state for constructing their stretch of the Ohio and Erie Canal. This magnificent sum allowed the Zoarites to pay off their mortgage by the 1830 deadline, and have enough cash in reserve to em-

Looking north up the sidecut.

bark upon a grand scheme of improvements. It is interesting to note that the law of celebacy was done away with in 1829.

The Society of Separatists now embarked upon a system of capitalistic communism that was as successful as it was unique. The old mill race was enlarged and a guardlock constructed at its head in 1830. The State of Ohio and the community of Zoar jointly financed the building of a guardlock in the feeder canal on the west side of the river. This now meant that canal boats could leave the Ohio and Erie Canal through the feeder guardlock, cross the slackwater pool in the Tuscarawas, then enter the sidecut on the east side of the river through that guardlock and service the grist mill and Zoar village.

A magnificent 80-room hotel was built within the village in 1833. Also in 1833, the Zoarites constructed a large general store across the street from the new hotel. Here, goods produced by the society members were sold to outsiders. In 1835, another hotel was constructed on a bluff overlooking the west bank of the canal. This new hotel catered to the canal passenger packets, with good overnight accommodations and great food.

An iron foundry was built to the west of the canal and a short branch canal was constructed to connect with the main waterway below the second lock (number 8) of the four south of Bolivar known to boatmen as the "Zoar locks." Canal boats were then able to carry pig iron to market, both north and south on the main canal. The society at one time owned at least four canal boats. The council was not above hiring outsiders to set up some of these new industries. The first overseer of the new iron foundry was an outsider, as were the first canal boat captains. Later, Zoarites were trained and placed in these positions. A new, magnificent mill, six stories high, was built south of the "Zoar Mile Lock" (number 10) in 1837. It straddled the canal. Boats actually entered this structure to load and unload.

It is speculated that by about this time the sidecut canal to the old mill south of town was no longer being used by canal boats. The old mill was no longer operational, so the always tedious, sometimes a bit dangerous, journey across the Tuscarawas was no longer required. The mill race was no doubt still in use, as a saw mill, woolen mill, and planing mill were in operation along its waters. It doesn't appear, though, that boats used it much after the new grist mill was built. The Society of Separatists continued to prosper, and it was necessary to refurbish and reopen the old mill south of town in 1848. Around 1878, a new mill dam, with a breast five feet above low water, was constructed near the site of the original dam. A permanent wooden gate was placed in the lower gate-support structure of the old sidecut guardlock, probably sometime prior to this. Four cast-iron wickets, or paddle valves, installed along the bottom of this permanent gate regulated water into the sidecut, which by this time was known only as the "mill race."

Ohio's Board of Public Works, the operating agency of the state canals, always working on a minimal budget, removed the upper gate-support structure of the feeder guardlock after it was no longer needed to pass canal boats, and used its stone in retaining walls as required along the system. The lower gate-support structure was used as a simple gate to regulate the flow of feed water into the canal. The state feeder dam continued to be a maintenance problem. After a while it was no longer maintained, the state preferring instead to pay the community of Zoar $100 per year to use the water backed up behind its new, higher dam.

While the Society of Separatists continued to prosper, with more than 7,000 acres of prime land and holdings valued at more than a million dollars just prior to the Civil War, little of this prosperity was being "wasted" on the society's members. Each member worked long and hard hours, at jobs assigned by the governing council. Members received a "living," while the fruits of their labor went to enrich the society.

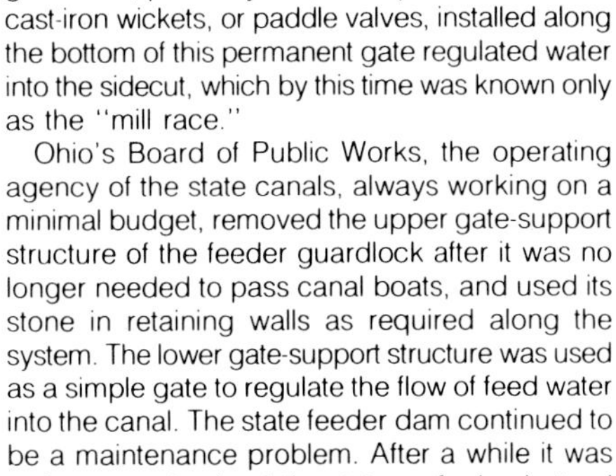

The east wall of the sidecut's lower gate-support structure.

Joseph Bimler, the leader of the Zoarites, died in 1852. Soon, the Civil War, railroads, and the example of the tourists relaxing in Zoar's hotels, brought pressures and temptations to the Zoarites' way of life that they were unable to withstand without the strong leadership Bimler had provided. Many of the fruits of the craftspersons' and farmers' hands now began to find their way to outsiders through private sales, rather than being offered through the society for the common good. Finally, in 1899, the Society of Separatists was disbanded. Each of the members received his or her rightful share of the communal property. The old mill south of the village was acquired by the miller at that time, and he continued to operate it into the 1920s. It was moved to the side of route 800 during the 1940s and operated as a store for tourists.

The state attempted to rebuild the Ohio and Erie Canal during the early 1900s. All the locks built by the Zoarites in the 1820s and 1830s were dismantled and rebuilt with concrete. The only exceptions to this were the "tumbles" (falls on the bypass channels) at locks 7 and 8, the lower gate-support structure to the feeder guardlock, and the guardlock to the sidecut on the east bank of the river.

The state acquired the Zoar mill race dam and covered it with a "permanent" concrete cap in 1908. It also built a 200-foot extension to the west and tied the western abutment of the new extension to the embankment of the Wheeling & Lake Erie Railroad with a 200-foot-long, 8-foot-high levee.

The Zoar mill, erected in 1821, refurbished in 1845, used until the early 1920s, photographed in the 1930s.

The reconstruction was never completed. The state legislature had a change of heart and authorized no more funds for the project after 1910. Then, in March of 1913, a flood of unprecedented destructive force tore out many of the new concrete structures in the northern part of the canal. They were never rebuilt. The new Zoar dam, however, withstood the forces of the flood. The stretch of canal south to the Sugar Creek Dam at Dover, which didn't hold, was one of the few sections of canal that contained water after the 1913 flood. Subsequent floods, however, did what the "Big One" couldn't—breached the earthen levee at the western abutment and swept away about half of the 200-foot extension.

The village of Zoar is now, largely, a tourist attraction, administered by the Ohio Historical Society. A 25-foot-high earthen dike along the western and southern limits of the village, constructed during the 1930s to protect Zoar from the backwater of a flood-control dam built then in the valley, have effectively isolated the remains of the feeder dam and sidecut guardlock from the main village. Efforts by several organizations are now under way to stabilize and preserve these canal structures, and perhaps make them once again part of the interesting and unique story of Zoar.

NEW BOOK ON ALEXANDRIA CANAL

During the Civil War, boat traffic on the Alexandria Canal stopped and a bridge was built across the Tide Lock (left). This famous Brady photograph was our only view of the lock until it was excavated in 1979 and 1985. (U.S. Signal Corps photo, National Archives)

The Alexandria Canal: Its History and Preservation, by Thomas Swiftwater Hahn and Emory L. Kemp, West Virginia University Press, 1992.

Reviewed by Bill Trout

Canal buffs will be delighted to know that the very first monograph published by West Virginia University's Institute for the History of Technology & Industrial Archaeology, is on a canal. And as a Virginian, I am especially pleased that this a Virginian canal—it branched off the Chesapeake & Ohio Canal but it was built by a separate company in 1843 to take coal freighters from Georgetown, now part of Washington, D.C., down to Alexandria's deepwater port on the Potomac.

The Alexandria Canal was only seven miles long, but it had all of the engineering features a notable canal requires, including a flight of four locks, a stone-arch aqueduct over Four-Mile Run, and a 1100-foot aqueduct across the Potomac. Unfortunately for the canal, the same progress which was responsible for its construction, has been responsible for eliminating nearly all traces of it in this solidly developed area surrounding the nation's capital. In fact, there is some debate about the exact route of the canal in relation to modern buildings. Could one of the secret rooms under the Pentagon be in the old canal bed?

The canal's most famous structure, the Alexandria Canal Aqueduct, was almost completely demolished in 1962, the victim of a foolish law in our land requiring that old bridge piers be removed because they "obstruct navigation," even when there is no longer any navigation to be obstructed. Therefore it takes a campaign to save a historic bridge or an aqueduct, and it took the Arlington Historical Society a lot of work to save one pier on the Virginia side.

The flight of locks, too, was almost lost forever. Over the years the locks became completely buried and lost to view. Even the ownership of the lowest lock in the flight, the Tide Lock, was in dispute because the federal government claimed all the land up to the high tide mark in 1776. Two centuries later when developers wanted to use the site, the federal claim was part of the leverage which led to a compromise—the development of the site, but the uncovering and preservation of the lock as a waterfront feature, and the establishment of the Alexandria Waterfront Museum beside it.

The first excavation at the lock was by Vivienne Mitchell and Pamela Gressey in 1979, to prove that the lock was still there to preserve and to place on the National Register of Historic Places. Then in 1982 and 1985 Tom Hahn and Emory Kemp systematically excavated the lock for interpretation and restoration. It should be pointed out that their restoration recommendations were not entirely accepted by the developers. The lock is a static display today, without working gates, to avoid the "dirty Potomac water." Someday we hope to see the lock brought back to life so historic small craft can be berthed on display in front of the Waterfront Museum.

The Alexandria Canal is a 76-page, well-illustrated account beginning with the history of the Alexandria Canal (with a good introductory chapter on its place in canal development in Virginia and the world) and ending with the authors' archaeological study of the Tide Lock. It even has a glossary of canal terms, and extracts from the notebooks of the canal's engineer, Maskell C. Ewing. There are precious few books available on canal archaeology in this country so this important book is a must for every canaller's shelf. It is also a welcome change from useless contract archaeology reports made by investigators who may be good archaeologists, but who don't know a breast wall from a stop-gate slot. As anyone reading *American Canals* knows, Dr. Tom Hahn is a world-class canal archaeologist, one of the founders of the American Canal Society twenty years ago. And Dr. Kemp, who started the new Institute for the History of Technology and Industrial Archaeology, in 1989, is also a member of ACS and knows his canals. It ought to be a law that contract archaeologists doing a canal site, should be required to include a recognized canal archaeologist or a canal expert as an integral part of their team. Canal archaeology is a specialty, and it takes special knowledge to pose the relevant questions of use to canal historians. Fortunately, Tom Hahn and Emory Kemp know what they are talking about. Let this book be a good example for future work in the field of canal archaeology.

The Alexandria Canal is available at $15 postpaid (plus 84¢ tax in WV) from the American Canal & Transportation Center, P.O. Box 310, Shepherdstown, WV 25443. If you're a member of the C&O Canal Association or the Virginia Canals & Navigations Society, say so and you'll get $2 off until May 1st. Copies can also be purchased over the counter at the Alexandria Waterfront Museum at the foot of Montgomery Street in Alexandria, beside the Alexandria Tide Lock.

Other canal-related monographs in the works which are tentatively to be published by the WVU Institute of Technology and Industrial Archaeology, and distributed by the American Canal and Transportation Center in WV, are Hahn and Kemp's *The Natural Cement Industry of the Upper Potomac River Valley*, scheduled for Spring 1993; Hahn's *The Lock-houses and Lock-keepers of the Chesapeake* and *Ohio Canal;* and *A Glossary of North American and British Canal Terms*, an American Canal Society project by John Droege and his Glossary Committee.

DEFIANCE: FOCAL POINT OF OHIO HISTORY

by Linn Loomis

Defiance was one of the most vital communities on the Miami and Erie Canal between Dolphos and Toledo, and marks the confluence of the Tiffen and Auglaize Rivers with the Maumee. Pontiac Park, situated along Route 424 in the northeast portion of the city, and Tecumseh Park, on highway 111 to the south, show that the area was rich in Native American history long before the concept of Ohio's canals became reality. Between the death of Pontiac in 1769 and that of Tecumseh in 1813, the famous American military leader, General "Mad" Anthony Wayne made his historic mark in the Maumee Valley, establishing Defiance as a focal point of pioneer history. Wayne gave the name Defiance to the site when he established a fort here in 1794. It was an act of defiance for a unit of the American army to be in this part of the Northwest Territory, then still dominated by the English and their Native American allies.

Immediately south of the Maumee, the canal remnants can still be seen along Perry Street. Near the corner of Third and Perry streets, the remains of Lock 37 can be seen. This structure has been partially covered by a parking lot. South of it, old factories and warehouses give evidence of what was once the canal route.

Miami and Erie Canal crossed Five Mile Creek, Township Road 153, south of Defiance.

Near the corner of Gorman and Perry streets are found the well-preserved remains of Lock 36. Originally built of wood in 1843, Lock 36 was reconstructed in concrete in 1907-08, and then abandoned in 1913 when this portion of the canal became a victim of the railroads aided by the 1913 flood.

Tecumseh Park, on Highway 111 South of Defiance.

Pontiac Park, on 424, east of Defiance.

Ruins of Lock 36, near Perry and Gorman Streets in Defiance.

Lock 37 from the north.

South of Defiance, near where Five Mile Creek empties into the Auglaize River, the canal builders constructed a culvert to carry the canal over a narrow part of the creek.[1] This single-arch structure, which now carries Auglaize Township Road 153 over Five Mile Creek, was one of the largest of its kind constructed during Ohio's canal-building era.

[1]*Editor's note. We leave it to John W. Droege, chairman of the A.C.S. ad hoc committee to develop a glossary of canal terms, in consultation with committee member William Dzombak, to determine whether this structure, when it was conducting the canal over the creek, was in fact a culvert or an aqueduct. See* American Canals, *no. 78 (August 1991), pp. 8-9.*

MUSKINGUM RIVER NAVIGATION SESQUICENTENNIAL

Muskingum River Parkway headquarters hides among the trees at Lock #10

When the Ohio-and-Erie Canal was finally completed in 1832, and the Ohio River system was navigationally connected to the Great Lakes, the canal commissioners of Ohio may have felt that they could relax, at least momentarily, and contemplate the impressive results of their efforts. It was not to be, however. Without a pause, citizens whose interests lay within hailing distance of the canal began pressing for the construction of branches which would connect them with what it appeared would soon be the mainstream of commercial activity in the state. Among those most difficult to ignore were the residents of the Muskingum Valley. This was one of the earliest areas of settlement in Ohio, and remained one of the most prosperous, with substantial marketable surpluses of grain, tobacco, linseed oil, and other agricultural products, and bracketed between the busy processing and commercial centers of Marietta, at the confluence of the Muskingum and the Ohio, and Zanesville, at the Muskingum's approximate head of navigation.

Side Cut to Dresden

Even before the canal was open to navigation over its entire length, the politically potent Muskingum interests had successfully petitioned for a side-cut to Dresden, a Muskingum River town 16 miles above Zanesville. The effect of this, evidently unforeseen, was to divert upper Muskingum Valley produce from Zanesville to Cleveland. The next petition, therefore, was for the canalization of the river between Dresden and Zanesville.

The new waterway prospered for a time, but the development of alternative means of transportation brought declining use and dwindling revenues, and the consequent postponement of needed maintenance. By 1886, the people of Ohio and their government were happy to be able to unload the whole burden onto the broad shoulders of the army's Corps of Engineers, which, following the constitutional revolution wrought by the Civil War, was rapidly assuming jurisdiction over most of the nation's navigable waterways. The Corps attended to the neglected maintenance needs, later added an additional lock and dam (number 11, completed in 1910) above Zanesville, and still later removed lock and dam number 1, at Marietta, when the completion of the Belleville lock and dam on the Ohio River in 1969 made them redundant. Otherwise, the system as it exists today remains much as it was when it was built a century and a half ago.

There was a brief revival of commercial traffic during the period of federal administration, due to the development of coal resources in the Muskingum Valley. The coal ran out during the 1940s, however, and by 1950 there was nothing left but a small amount of upbound sand and gravel on the lower river. The Corps closed locks 2 through 11 in 1952, and lock 1 in 1954, and began to prepare for the demolition of all the structures, allowing the Muskingum to return to its natural state as a free-flowing river.

Recreational boating did not figure into the calculations of the Corps of Engineers, but it is important on the Muskingum. The valley is still densely populated and relatively prosperous. Many riverbank residents are boat owners, and they still know how to get a hearing in the state capital. The result is that the locks and dams were not demolished and instead were returned to state jurisdiction and reopened to traffic as the Muskingum River Parkway in 1958.

For many years, the State of Ohio operated but did not maintain the navigational facilities, and dilapidation became acute. Over the past four years, however, a major restoration project has been carried out. One consequence has been that although most of the locks were open to local traffic, it was impossible to cruise the waterway in its entirety because one or more of the locks was

The Marietta College crew works out on the Muskingum River.

The youngster of the Muskingum navigation system, Dam #11, built in 1910.

The showboat Becky Thatcher in port at Marietta.

always shut down for repair. Now, however, in 1991, the sesquicentennial year, all the locks will again be open. At the present time, they are operated routinely Friday through Sunday, and by request, on 3-hours notice, Monday through Thursday. Boaters planning a visit should check with the Parkway office for current information (P.O. Box 2806, Zanesville 43702; phone 614 452 3820).

Special events are planned throughout the 1991 boating season. These include living history displays, sternwheeler parades, and workshops dealing with natural history, folklore, and other topics. Special demonstration tours of the hand-operated locks will be conducted on Sundays through July and August at locks 8 and 10. The annual Zane's Trace Commemoration, June 14 through 16, will be dedicated to the Muskingum sesquicentennial this year. The year of celebration will culminate in a gala festival weekend September 14 and 15, with an antique boat show and many other events—the full schedule is not yet available. —DFR

Before restoration, some of the lock gates could barely hold water. This case in point is at Lock #10, Zanesville.

Marietta, the first permanent settlement in the Northwest Territory, abounds in historic buildings.

OLD SANTEE CANAL SPECIAL EVENTS

The Old Santee Canal State Park has issued a schedule of special events for the period July 21 through December 21, mostly on weekends. Those planning to visit the park, especially with children, might wish to obtain a schedule, as preregistration is required. For details, write the park at 900 Stoney Landing Rd., Moncks Corner, SC 29461, or phone 803 899 5200.

MAJOR REHABILITATION ON THE SCHUYLKILL

The two-and-a-half mile segment of the Schuylkill Canal known as the Oakes Reach (in honor of canal engineer Thomas Oakes) is one of the only two that have survived since the canal fell into disuse. The other is the Mont Clare/Port Providence Reach. Both are located in the Mont Clare, Pennsylvania vicinity, in Upper Providence Township, Montgomery County. Both have benefited from the coöperation of the township government, the state Department of Environmental Resources, the Pennsylvania Electric Company, the Schuylkill Canal Advisory Committee, and the Friends of the Schuylkill Canal, in promoting a number of restoration and preservation projects. The most recent of these, on the Oakes Reach, has also drawn in the legislature, through the efforts of state senator Edwin G. Holl. A $150,000 appropriation along with funds already in the Department of Environmental Resources budget will make possible the fortifying and stabilizing of the 147-foot stone guard wall above Lock 60, which has deteriorated to the point of being undermined. Work is scheduled to begin this spring.

[Information provided by the Pennsylvania Canal Society.]

DELAWARE CANAL STATE PARK

The Delaware Canal, connecting Bristol and Easton, Pennsylvania, became the Theodore Roosevelt State Park in 1940 and the Delaware Canal State Park in 1989. It has just been authorized to undertake a $2.2 million capital improvement project. The work will include rebuilding the Durham aqueduct and reconstructing a bridge over the canal. The park recently purchased the old lock house at New Hope, and is in the process of refurbishing it as an interpretive center and headquarters for the Friends of the Delaware Canal. Future plans call for the rebuilding of Lock 19 at Lodi.

[Information provided by the Pennsylvania Canal Society.]

CANAL DAYS AT PORT COLBORNE

The Port Colborne Historical and Marine Museum will celebrate "Canal Days" the 3rd and 4th of August, 1991. Located at the Lake Erie terminus of the Welland Canal, the museum is comprised of seven historic buildings furnished with artifacts of the area and its marine heritage. Canal Days is a festival celebrating the maritime character of Port Colborne and the Welland Canal. Arts and crafts, model boat and other marine displays and demonstrations, and food and entertainment are featured. The museum is located at 280 King Street, one mile south of highway 3. Hours are 10 to 5 on Saturday and 12 to 5 on Sunday; admission is $2, seniors and students $1.

The museum is open May through December, 12 to 5 daily, admission free except for special events. Afternoon tea, featuring homemade biscuits and jam, is a tradition from 2 to 4 every day but Monday, June through September. To arrange group tours or for further information, call 416 834 7604, or write the museum at P.O. Box 572, Port Colborne, Ontario L3K 5X8, Canada.

[Information provided by the Port Colborne Historical and Marine Museum.]

THE KEWEENAW WATERWAY, THEN AND NOW

The Keweenaw Waterway and the city of Houghton as they appear today. Photograph by Tom Duchaine of Keweenaw Graphic Images.

Michigan's Upper Peninsula is well known among canal enthusiasts as the site of the St. Marys Falls Canal, or Soo Canal, whose huge locks open Lake Superior to the shipping of the other Great Lakes, the St. Lawrence Seaway, and the world. There is another Upper Peninsula canal, however, of the same vintage as the Soo and, like the Soo, still in operation. This is the Keweenaw Waterway, which cuts across the Keweenaw Peninsula, the center of Michigan's copper country, where once "nuggets" of pure, metallic copper weighing from four to eight tons each were dug out of the ground. The Keweenaw Peninsula projects up into Lake Superior in a northeasterly direction; the Keweenaw Waterway cuts across it near its base. It appears to be a short-cut from one place to another but those are not the places between which shipping ordinarily goes. Shipping between Marquette, Michigan, and ports in Wisconsin and Minnesota can save up to about 25 miles using the waterway, but between the Soo and Duluth it is five miles longer than the open lake route. According to the Coast Guard, its main use by commercial vessels is as a harbor of refuge during the stormy months of October and November. It is a flat canal, like the Cape Cod and the Suez, so there are no locks to be maintained and operated. This probably helps to explain why it remains open despite its near redundancy.

It was a very different story, however, when the Keweenaw Waterway first became available as a shipping route. We have recently come across an account of the opening of one end of the waterway, and the impact of this development on the region. Any of our readers who may be familiar with the history of the other end are invited to contribute the rest of the story. The present waterway as a single unit dates from the federal government's purchase in 1891 of two canals: the Portage Lake and River Improvement Company Canal, from Portage Lake southeast to Keweenaw Bay; and the Lake Superior Ship Canal Railway and Iron Company Canal (a name only slightly shorter than the canal itself) from Portage Lake northwest to Lake Superior. It was the Portage Lake and River canal that was built first, and it is of its inauguration that we have an account.

The Portage River Improvement Company, as it was originally known, was formed in 1859 by some of the copper mining companies and two leading citizens of the copper country, Ransom Sheldon and his brother-in-law, C.C. Douglass. They hired a contractor, W.W. Williams of New York, and an engineer, John Harris Forster. Fortunately for us, Forster was a prolific writer. He kept journals, wrote articles for engineering journals, and contributed to the proceedings of the Michigan Pioneer and Historical Society. It was at the annual meeting of the latter group, in June 1885, that he discussed the opening of the new canal that was to become the lower half of the Keweenaw Waterway.

"Previous to this [Forster wrote] all steamers from the lower lakes came to anchor in the roadstead off the mouth of Portage river. Freight was discharged into scows, which were towed into the river and up to the mines by small tugs. During the season of storms, several days were sometimes required to discharge a steamer, for, if a storm arose, the steamer was obliged to cut and run for L'Anse for shelter, returning after the storm had subsided. The cost, delay, and damage occasioned by this transfer was enormous and became intolerable to businessmen. Hence came the Portage River Improvement. Passengers were

(Concluded on Page 51)

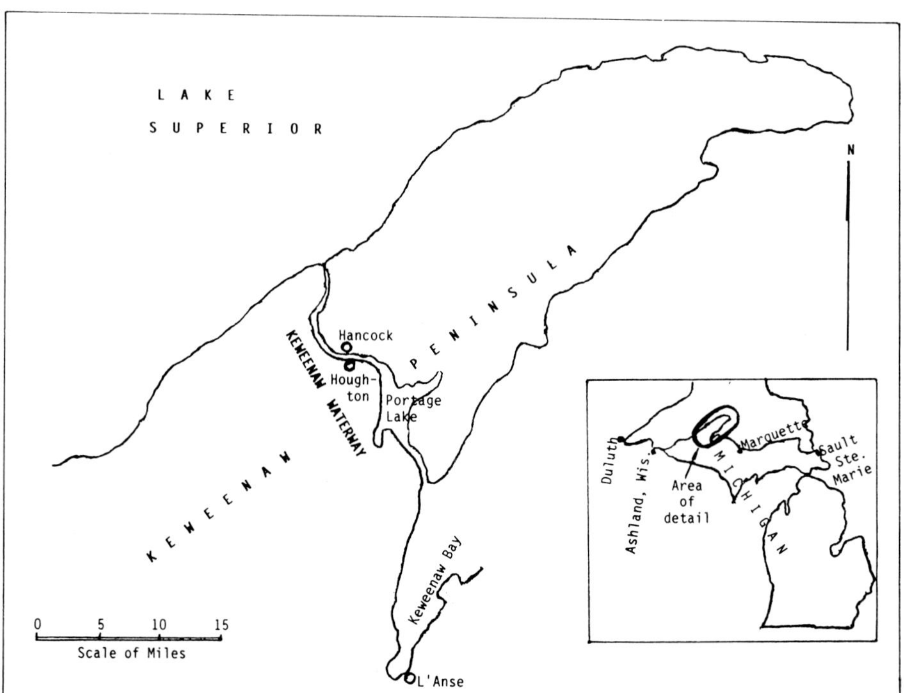

THE KEWEENAW WATERWAY

(Concluded from Page 50)

subject to like detentions. There were no means of communicating with the entry only by small boat or tug. On land a dense wilderness of trees and swamps intervened. The date of the arrival or departure of steamers from the lower lakes was uncertain. So the Houghton man had to go at a venture; when he arrived at the entry there was often no boat, or one had just departed. He was in for several days' waiting. His only shelter was a log house, his food of the roughest kind, his bed a pair of blankets spread on a floor, in the midst of filth and vermin. The mosquitoes and black flies tormented him day and night. His only amusement was playing cards and drinking bad whiskey with his unfortunate fellow travelers."

Small wonder, then, that the opening of the canal was an occasion for celebration. The first ship, a small steamer named "Detroit," and originating in that city, arrived at the wharves of Houghton and Hancock in June 1860. "Great was the rejoicing of the inhabitants; fierce and loud the shrieking of stamp mills whistles, as the boat hove in sight. It was a gala day at Portage; a new era was inaugurated. Mr. S.L. Smith, now of Lansing, made the speech of welcome."

Fortunately or otherwise, we do not have the text of the speech of welcome delivered by Mr. S.L. Smith. Since he went on to Lansing, the state capital, it may be assumed that it was eloquent and stirring. The passing of the author and his words into obscurity testifies not to their inferiority but to the evanescence of even the worthiest of human efforts. In the case of the engineer Forster, however, both the words and the works have survived. The copper is long gone, except for some mining of the slag piles, and the cities of Houghton and Hancock no longer depend for their prosperity upon the commerce carried from Detroit and elsewhere via the Keweenaw Waterway. Tourism is now the leading industry of the Upper Peninsula, and recreational boating, the principal summertime use of the waterway, is only a minor component of tourism. During the autumnal "season of storms," however, it is still necessary for Lake Superior shipping "to cut and run . . . for shelter." The rejoicing then at the existence of the Keweenaw may not be as public as it was in June 1860, but it is surely just as heartfelt.

—D.F.R.

MARGUERITE II WORK UNDERWAY

From the CSO Newsletter July, 1992

From October 4 to 18 Ohio will witness one of the more interesting historical reenactments ever staged in the state when the Delphos Canal Commission undertakes the voyage of the *Marguerite II* from Toledo to the Tall Stacks 92 in Cincinnati.

The boat will be a true canal boat, modified to meet contemporary transportation standards. The boat is to be a recreation with much lumber from period sources using oak, ash, locust and walnut in many places. Among her deck boards will be an actual plank from the original *Marguerite,* a passenger packet which sank in Delphos. The remains of the original boat were recovered in 1987 by boys and girls from Delphos.

Construction of the timber boat has begun and is being built in Delphos.

A travel crew will accompany the new boat to not less than seven original canal cities, stopping to 'dock' the boat for local tours, especially for school children. Among the stops on the voyage will be International Park in Toledo, Carillon Park in Dayton and Sawyer Point in Cincinnati. Defiance, St. Marys, Piqua and Middletown/Lockland will also be sites for stops.

The Delphos Canal Commission anticipates that up to a half million people will see the boat during its four day stay at the Tall Stacks 92 event in October in Cincinnati.

AMERICAN PETROLEUM INDUSTRY BORN IN PENNSYLVANIA CANAL

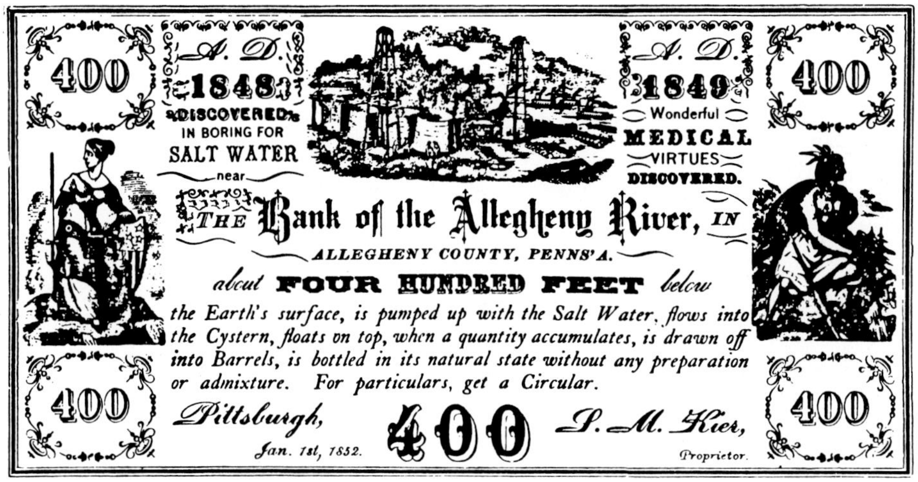

A whimsical advertisement for petroleum as medicine.

By William Dzombak

It is generally recognized that the use of petroleum in America, and then throughout the world, began in 1859, when Edwin L. Drake drilled an exploratory well at Titusville, Pennsylvania, that released a gusher of oil. The world was changed by that event. Drake's well inaugurated the large-scale production of petroleum products used first for illumination (kerosene lamps), next for lubrication of steam engines, and later as fuel for automobiles.

The petroleum products needed for these purposes were supplied by an oil industry that was launched by Drake's well, but there would not have been a Drake well, to probe for oil, if there had not already existed a recognition of the potential uses of petroleum. It is not generally known that the useful properties of petroleum were first demonstrated in a consequential way by a dramatic event that occured in 1845 on the Pennsylvania Main Line Canal at Tarentum, near Pittsburgh.

On that momentous day, some boys playing near the canal happened to toss a piece of burning wood into the water in the canal. To their surprise, the canal burst into flames! A section of canal at least one-half mile long was brilliantly ablaze, under dense clouds of smoke! Everyone who witnessed the event was astonished to see fire on the water—the discovery that petroleum would burn, and give off light and heat.

Everyone realized, immediately, why the canal was burning: the surface of the canal was covered with petroleum discarded into the canal from nearby salt wells. Brine obtained from the wells was accompanied by a dark fluid, foul smelling, that contaminated the salt produced by evaporation of the brine. To obtain pure salt, brine from the wells was first pumped into a vat, where the nuisance "rock oil" floated to the surface and then spilled over the rim of the vat and trickled away, ending up in the canal adjacent to the saltworks. The firebrand thrown by the boys had ignited the oil floating on the canal.

Before the great canal fire at Tarentum, few people had ever heard of petroleum. Soon after, people were using "rock oil" lamps to illuminate their homes and barns. Oil-burning lamps were soon attached to the caps worn by coal miners as they dug fuel for the furnace fires that heated evaporating pans at the saltworks located along the canal. The canal fire delivered light into dark homes, and granted people a later bedtime, for "rock oil" was given away free, while lard for lamps and tallow for candles were both too expensive to be used merely for reading or other leisure pursuits after a day's work was done. The people of Tarentum, who were thus profoundly affected by the introduction of petroleum into their lives, memorialized the epoch-making fire and thereafter compared living conditions "before the canal fire" with those "after the canal fire."

Within a few years, oil-burning lamps were being marketed by Samuel M. Kier, who operated one of the wells and saltworks near Tarentum. Kier also promoted the medicinal value of petroleum, taken internally or applied externally, and shipped bottles of "Kier's Rock Oil" as far as the gold fields of California. Samuel Kier's principal source of income, however, was the canal business, for he was one of the chief owners of the Mechanic's Line of canal boats, which operated between Pittsburgh and Philadelphia, with a branch-line operation to tidewater at Havre de Grace, Maryland. Kier was also the first person to refine petroleum by distillation, in an effort to remove offensive odors from the product that was to be burned in lamps or sold as medicine. Kier's pioneer work on petroleum refining was performed near his canal office in Pittsburgh.

The connection of the Pennsylvania canal with the birth of the petroleum industry is thus a double link: a serendipitous canal fire resulting from the leakage of "rock oil" from a saltworks brine well operated by Samuel Kier, owner of a line of canal boats. It is worth noting, also, that the first oil well, at Titusville, was drilled by "Uncle Billy" Smith, a salt-well driller recruited by Colonel Drake from the saltworks at Tarentum soon after the canal fire directed the attention of the world to the valuable properties of petroleum.

CANAL GHOSTS ON THE OHIO & ERIE

By Terry K. Woods

I think perhaps all 19th Century Americans believed in ghosts, if only to relieve the monotony of their rather dull, uneventful lives. Though most 'Town Jakes' assumed a canal boatmen's life was more eventful and exciting than theirs, (and boatmen all agreed), it wasn't really all that exciting. Boatmen believe in ghosts, too. Ghosts were just something a boatman might run into anytime. Listen to a story Wick Ludenberger, driver of his Dad's boat, BOLIVAR, in 1904, told us some years ago, . . .

"There was just the Old Man and me one night. We was runnin' light come'n from the Papermill in Akron an' we was try'n ta get back down ta Bolivar before we laid over, so we was runnen after dark. Well, I was out there on the towpath. An we had a light on the bow, just a big Kerosene lantern with a reflector on it. Didn't cast no more of a beam than ya could spit. And all of a sudden I saw this big white thing loomin' on the towpath as we was com'n round a bend. Well, GHOST, I thought. I was sure it was a ghost. The Old man, he didn't do nuthin. I was driv'n' Big Jack, (ya could pull a light boat with one horse, ya know.) Well, he kept plodd'n' along as if there was nuthin there. I don't know if horses can see ghosts or not.

"So I got up on Big Jack and sort of hunkered down behind his head. I figured if horses couldn't see ghosts, maybe ghosts couldn't see horses. I was all shiver'n' and shak'n' as we got closer and closer and - this ghost got bigger, stood up, and — mooed! It was a damn, big white cow! It had been lay'n' on the towpath an we'd disturbed her. I was so relieved. We didn't see another ghost that whole trip."

Ghosts were just part of the natural order of things. Wick's brother Ben told me one day that Lock #4 was haunted. When asked why, he paused, looked puzzled and answered, "I don't know. Maybe somebody saw a ghost there once." So Lock #4, south of Canal Fulton was haunted — not so much the lock as an old frame building that stood near the lock when the boys boated with their Father and Mother in the late 90's. Neither of the boys actually ever saw a ghost there, but they'd heard enough stories to ensure that they wouldn't be found anywhere near Lock #4 after dark.

Lonesome Lock (Lock #29 north of Akron), a good two miles by canal through desolate country from Peninsula, was also haunted. In fact, some of the older boatmen called it "Haunted Lock". It wasn't so much that ghosts had been seen there as that it just should have been haunted. Everyone knew the stories of how, in the 'olden times' robbers had laid in wait for unsuspecting crews here at this lock, without so much as a tavern to provide support, to set upon them, rob them, murder them, and throw their mangled bodies into the canal. Of course, no one you ever met had actually seen any of this carnage. It had all happened long ago. But you knew that whole area was haunted. You practically held your breath until your boat had passed this spot and was near Boston Mills, and civilization.

A more conventional ghost story concerns 'Amy' and an old white brick tavern on the berm bank of the canal some eight or ten miles south of Roscoe. Shortly after the finish of the Civil War, 'Amy', (ghosts never seem to have last names) left Cleveland in a canal boat. She was to meet her fiance, a Captain in the Union Army who had

just been released from a Confederate prison, at this tavern. They were then to be married, go out west somewhere, and live happily ever after. Unfortunately for 'Amy', and fortunately for the story, she was followed by a rejected suitor. He found her at the window of an upper room in the tavern, gazing off down the canal hoping to catch first sight of her lover's craft as it brought him to her. The crazed suitor sneaked up behind 'Amy', and, reaching around her, slit her throat with a large knife he carried. She died there, her blood staining the window sill, her eyes still staring down the canal.

It isn't clear what happened when Amy's' fiance arrived and found her dead. Some accounts have him chasing her murderer north and killing him on sight. Others have him wandering off in the moonlight, driven mad at the sight of his murdered loved one. What happened to the fiance isn't important. What is important, is that 'Amy' is still at the tavern, as a ghost. She inhabits that front, second story room. Many people have seen her face at the window, still gazing off down the canal. Many people have heard her in that room, smashing things. (She has a temper, for a ghost.) No one could stay in that room without sensing her presence, and having his (or her) belongings strewn about the room. Much later, in the 1970's when that old tavern briefly became a restaurant, the owner showed me the 'bloodstains' still on the wooden sill. And I talked with waitresses who, after a few sessions of spilled food and smashed dishes, refused to serve any more meals in that room. 'Amy', it seems, is a spiteful ghost. It's hard to say what became of 'Amy'. The Tavern is vacant again — has been for years. 'Amy' is a hard ghost to live with. I can't help but wonder if she was any easier to live with when alive, and whether her fiance might have had the better of it after all.

All ghosts, it seems, were stay-at-homes, unable or unwilling to leave their 'haunt'. This proved fatal to a group, no one knows how many, who haunted a cluster of abandoned canal boats in a basin just south of Barberton after traffic had come almost to a standstill. Local residents were annoyed and awakened nearly every night by the shouting, screaming, and blood-curdling yells coming from these boats. But when the neighbors and police investigated, no matter how quickly they responded, there was never anyone to be found — nor any evidence that human beings were at all involved! Obviously, "prankish ghosts"! And the pranks continued. One group of local residents finally had enough. One dark night, when the ghosts were particularly noisy and objectionable, these residents stealthily approached the abandoned craft by rowboat, and set everything in the basin afire. From the safety of the far shore, the residents watched in fascination as the flames leaped higher and higher, consuming everything; and listened in growing horror to the groans, screams, and finally whimpering, as the lashing flames consumed every boat in the basin to the waterline. Rather drastic measures perhaps, but the local residents were no longer kept awake nights by rowdy 'ghosts'.

These are a few 'canal ghosts' stories from my collection. Can any of the readers of *AMERICAN CANALS* top these?

TRIALS AND TRIBULATIONS

By Bill McKelvey

The following is excerpted from the 1868 proceedings of the New York State Court of Appeals Under the heading "John E. Stone versus the Western Transportation Company, Opinion of the Court, per Mason, J."

Mason J. This action was brought to recover the value of a quantity of hay and oats, some wagons, harnesses, farming utensils, etc. This property, at the time it was destroyed by fire, was in a barn in the possession of the plaintiff, in which a number of horses belonging to the defendant were being wintered, under a written contract between the plaintiff and the defendant, in which the plaintiff agreed to keep twenty horses for the defendant through the winter, and to furnish and supply said horses with good and commodious stables and yard room, and to furnish an abundance of hay for food, and salt for use of the horses, and to furnish sufficient straw for bedding, and to keep constantly a sufficient supply of water conveniently situated for the use of said horses, and to furnish lamps, oil, and all tools which may be requisite or necessary in grooming or taking care of said horses, and to furnish a man and team to assist in leading the said horses to the tow-path of the Erie canal in the spring, or when said horses may be required by said company. The plaintiff also agreed to board a man for the said defendant at two dollars per week, in case the defendant may require a man to take care of said horses. I think it pretty clear, taking this contract as a whole, that it contemplated the defendant should furnish a man to take care of the horses.

It is true the plaintiff agreed to keep these horses for ninety cents per week for each horse. But here are the positive undertakings specifically provided, to furnish hay for food and straw for bedding, and such quantities of salt as may from time to time be required for the use of said horses, and lamps, oil and tools which may be requisite or necessary in grooming or taking care of said horses, and to keep constantly a good and sufficient supply of water conveniently situated for the use of said horses.

There is no agreement to furnish a man to feed and groom the horses, but there is an agreement to board the defendant's man for two dollars per week, if the defendant should require it. And there is also an agreement to furnish a man and team to assist in sending the horses to the canal in the spring. The parties contemplated that the defendant would furnish a man to feed and take care of the horses. Be this as it may, however, the contract gave the defendants this right, to put the horses in charge of their own man to feed and groom, and the plaintiff agreed to board him for two dollars per week. This man must be deemed the servant of the defendants and not the plaintiff.

THE OHIO AND ERIE CANAL CORRIDOR COALITION

Northeastern Ohio is blessed with many remnants of a once prosperous canal system, which was started in 1825 and connected Cleveland on Lake Erie, with Portsmouth, on the Ohio River. Many structures remain due to a State effort to rebuild, as New York had done, in 1907, but this effort was cut short by the devastating flood of 1913. The Ohio and Erie Canal Corridor Coalition is a grassroots, local effort to preserve and promote usage of these remnants, extending from Cleveland to Zoar, and eventually down to the Ohio River.

The OECCC began with a public meeting on September 21, 1989 and has been expanding ever since. Immediately, a steering committee was set up of nine local volunteers who were supported by liaison with the National Park Service. A quarterly newsletter was created and was appropriately named "THE SILVER RIBBON" after an epic ballad by Captain Pearl R. Nye, a legendary, real Ohio Canal boat captain whose saga is known throughout the state. An inventory and public hike was held from Cleveland to Zoar and over 100 citizens attended who compiled local information to be used in the future; the canal was rediscovered.

Our Corridor is unique in that it stretches over 85 miles, and consists of two river systems, separated by the 41st Parallel. Cleveland is situated at the mouth of the Cuyahoga River and Lake Erie, from which it grew, and is followed by 22 miles of the blossoming Cuyahoga Valley National Recreation Area and its Valley Rail Line. The North Cuyahoga Valley Inc. is a citizen's group of volunteers who have been working together for over 10 years to show the resources of Cleveland; it is a diverse area consisting of interstate highways, international airports, the seaport of Cleveland, and many major manufacturing facilities, and yet was built along historic Indian trails and ancient transportation networks whose presence can still be found.

The Cuyahoga Valley National Recreation Area is just South of Cleveland and is connected by its newly constructed Towpath Trail Project. John Debo, CVNRA Park Superintendent, spearheaded this effort by getting cooperation from many local leaders, Cleveland and Akron Metro-Park Systems, and numerous individual communities affected by the Park. The trail is presently under construction and will span roughly twenty miles of continuous all-purpose surface, over aqueducts, boardwalks, and many other canal improvements, through tunnels and under railroads, and includes two beautiful visitor's centers and the newly reconstructed Lock 38 North, an operating Lock, with rangers and audio visual, high tech assistance. The Towpath Trail connects to the Buckeye Trail, the State of Ohio system of over 1,000 miles, and to the North Country Trail, a national system that will cover over 3200 miles of hiking network.

The canal from Akron to Zoar was supplied by the Tuscarawas River through the Portage Lakes Feeder System. Akron is represented by the Cascade Locks Association, an effort to preserve a staircase of 8 Locks in the heart of downtown, as well as the Mustil's Store, a canal era grocery, and its lock tender's house. Progress Through Preservation is another local group leading the way, whose efforts concentrate on preserving buildings, engineered structures, and other historic artifacts dear to the "Rubber City". City management has earmarked over $100,000 to the start of construction of the Cascade Trail, almost in the central business district. Summit County is cooperating with the extension of this effort, outside the city—it has utilized canal lands with a beautiful park system that goes back to the 1920's, and is finally being re-discovered.

There are over nine miles of navigable canal water reaching from Akron's Summit Lake to Barberton, Ohio. The "Magic City" is represented by the Wolf Creek Association and other local groups. Barberton has Lake Anna, in the heart of town, the Yoder brother's restored barn (now classy office buldings), remnants of 4 more O.C. Barber structures, Wolf Creek Lock (buried), miles of beautiful towpath, and of course, the Wolf Creek Aqueduct, the only known structure of its kind in the entire state. The canal is surrounded by industry, both past and present, and yet the solitude of the canal lands is undescribable.

Clinton was once destined to be the terminus of two canal systems, and it still retains much of its historic beauty. Many structures have not changed since canal days, and Clinton still has two beautiful locks as well as a guard lock, protecting the City, and many of its homes appear as they did over 100 years ago.

Canal Fulton has never given up its heritage, as can be seen from the St. Helena III, a modern concrete Canal boat which is operated each summer by the Canal Fulton Heritage Society, and their older boat the St. Helena II that carried thousands of passengers for over 15 years which they hope to turn into a museum, and use again. Canal Fulton has recently done much to revitalize its downtown area. It maintains the McLauglin Drydock facility, has miles of walkable towpath, and Lock 4 Park has been refurbished again by Stark County, with doors, a locktender's house, and an abandoned mill site.

Massillon grew due to the canal and the city is full of old stone buildings and churches. The river is beautiful as it runs past steel mills and under rail lines, and their many museums reflect the town's glorious past. Its close neighboring city, Canton, was the home of past President McKinley, and the recently refurbished McKinley Memorial is a stone structure comparable to any memorial in the World. Canton also has the Football Hall of Fame and many other modern attractions.

Navarre is a collection of three Canal towns: Bethlehem, Rochester, and Navarre, all of which were unique. It has established the Navarre-Bethlehem Township Canal Corridor Group of volunteers, and they are in the process of making a modern park around the canal. Stark County, always a canal supporter, has done its best to acquire most rights within its borders.

Tuscarawas County has Locks 7, 8, 9, and 10 within its boundaries, all of which are in good shape, and are being considered for restoration. The Canal town of Bolivar was located near two aqueducts, and is at the junction of the Sandy and Beaver Canal. It has many Canal era structures and is located near Fort Laurens, the only revolutionary fort in Ohio. Zoar Village was instrumental in Canal construction and many, many of its buildings are still as they were in Canal Days. The feeder dam is located here, as well as their guard Lock that still shows the 1830 date on its cornerstone. Zoarville and nearby Camp Tuscazoar also exhibit much history, as well as the Muskingum Water Conservation District of the 1930's.

Presently, a feasability study supported by Congressman Ralph Regula, a long-time friend of canal preservation, is being conducted by the National Park Service, and is expected to be submitted shortly. OECCC volunteers have developed a slide show, and have sponsored hikes, bike ventures, canoe races, river clean ups and area tours, and the Second International Conference on Historic Canals was recently held in Akron's Quaker Square. Canal videos have been made of the area, classes are held at the local universities on canal subjects, the Akron McDonald's is built along the Canal theme, and the "BIG 160", a canalabration commemorating the opening of the system was recently held with coordinated activities from Cleveland to Zoar.

The Ohio and Erie Canal links 85 miles of greenspace, connecting culture, history, and architecture, through common interests. The Ohio and Erie Canal Corridor Coalition is working steadily at becoming the fourth National Heritage Corridor. Much progress has occurred, cooperation has been excellent, and our future looks bright.

Larry Turner, 11619 Fraze Road, Doylestown, OH 44230.

Donald L. Morris of Point Pleasant (Bucks County) PA sent us the above photo of the Point Pleasant Aqueduct carrying water of the Delaware Canal over Tohickon Creek. The original aqueduct, thought to be cast iron and wood, was replaced in concrete in the 1940's. The Delaware Canal is one of the few historic canals in the USA which (except for a few breaks in the canal banks) has been carrying water continuously since it was built in 1832, for nearly its entire 60 miles. The people of Delaware Valley are extremely protective of their canal, and any intrusion of highways or bridges which threaten it.

THE ARKANSAS TRAVELER AFLOAT

By David F. Ross

Above Little Rock, Ozark mountain scenery takes over.

The political dimension of American waterway development is frequently commemorated in the naming (and renaming) of locks and dams. Just in the recent past, for example, such nonpolitically designated structures as Aliceville L&D (on the Tenn-Tom), Warrior L&D (on the Black Warrior), L&D 26 (on the Mississippi) and Gallipolis L&D (on the Ohio) have been renamed to honor politicians from their respective areas. There is probably no other agency of the executive branch which has been as successful as the Corps of Engineers in developing a symbiotic relationship with Congress, and the large number of Corps-built-and-administered facilities named to honor politicians is an apt expression of this intimacy. The waterways themselves, however, have largely remained outwardly aloof from the political influence. Wags may have made reference to "Clinton's Ditch," but it was officially the Erie Canal, and that precedent of employing geographically informative rather than personally honorary names for waterways has been followed almost without exception, from the Ohio and Erie, through the Panama, down to the Tennessee-Tombigbee. Even the St. Marys canal is named for the river, not for the lady.

Within the Corps of Engineers, technology, economics, and politics are so tightly interwoven that it is usually impossible to identify any particular action or recommendation as uniquely an expression of one or another. In those rare instances, however, where the Corps and Congress part company, with Congress voting to fund the construction of a waterway over an adverse recommendation from the Corps, we may be sure that the waterway in question is peculiarly political in nature. Even such waterways, however, tend to remain nomenclaturally chaste. An outstanding example is the canalized Upper Mississippi, which by rights should be the Herbert Hoover Waterway, but is not. The one exception to this rule is the McClellan-Kerr Arkansas River Navigation System, commonly but improperly and inaccurately known to boaters as the Arkansas River. (Publisher's Note: In view of the attention focused on Little Rock in the recent presidential election, would it be fair to rename this system: "The NEW Clinton's Ditch"??)

The Arkansas River was navigated by steamboats as early as 1820, with navigation reaching as far as Little Rock by 1822, and Fort Gibson, near the present site of Muskogee, Oklahoma, by 1827. Eventually, at least one steamboat (the snagboat *Wichita*) reached as far as 65 miles above Tulsa. Navigation was difficult and dangerous, however, because of shallow water except during seasonal floods, shifting channels due to heavy siltation, and snags. As elsewhere, river traffic declined rapidly as soon as a more reliable alternative became available through the expansion of the railway system. Catastrophic flooding, especially in 1833, 1844, 1876, 1877, 1912, 1927, and 1943, kept alive interest in river improvement on the Arkansas, but support for improvements which included navigational development was difficult to muster. Such as it was, it came mainly from industrial development promoters in the new state of Oklahoma (admitted to the Union in 1907), who conceived of cheap river transport as a locational attraction, and from farming interests in Arkansas, who perceived themselves as victims of a railroad monopoly.

Agricultural commodities constitute the principal freight carried on the Arkansas.

In 1928, the Corps was directed by Congress to undertake a study of the feasibility of bringing the Arkansas River under control. The study, completed in 1935, concluded that the annual operating costs of navigational improvements would exceed the benefits in freight cost savings by a ratio of almost 2 to 1. A second survey was begun in 1939 and completed in 1945. This span of years is significant because it brackets not only the 1943 flood but also the 1942 elections of Robert S. Kerr as Governor of Oklahoma and John L. McClellan as Senator from Arkansas. When the completed survey was forwarded to Congress in 1945, it carried a mixed endorsement from the Board of Engineers for Rivers and Harbors. The Board recommended that its flood-control proposals be effected immediately but that its navigational proposals be deferred indefinitely. Although the estimated cost-to-benefit ratio was considerably more favorable than that of the earlier study, it was still greater than 1. On this occasion, however, Congress voted to overrule the Board, and in 1946, President Truman signed into law a bill authorizing construction of the Arkansas River Navigation System.

Rock jetties are a common feature of the Arkansas. Sometimes submerged, they can be hazardous to the unwary.

As many advocates of civil works projects have learned to their sorrow, however, congressional authorization is a long way from the end of the struggle. Rather than meat on the table, it is only a hunting license: the real feast does not begin until Congress appropriates funds. In the case of the Arkansas, that took another 11 years. In the meantime, in 1948 Kerr joined McClellan in the Senate, and went on to become Chairman of the Public Works Subcommittee for Flood Control and Rivers and Harbors. The political waters that would eventually result in a nine-foot channel were slowly but implacably rising to a crest. Construction began in 1957, and at the end of 1970, the waterway was an operational reality.

At $1.2 billion, the McClellan-Kerr Arkansas River Navigation System was the most expensive project ever undertaken by the Corps of Engineers until it lost this honor to the $2 billion Tennessee-Tombigbee Waterway, completed in 1985. As an economist, I am aware that the results of

With a normal lift of 54 feet, Dardanelle Lock is the highest, and backs up the largest lake.

cost/benefit analyses depend overwhelmingly upon arbitrary assumptions, which in turn depend upon the nature of the result desired by the analyst. I therefore refrain from venturing an opinion as to whether the taxpayers were well or poorly served by those stewards of the public purse who acknowledged the leadership of Robert Kerr and John McClellen. Whether it honors or disgraces their memories, the cost has been sunk and the waterway is there. As the voting public, we need to worry about future river development projects; as the boating public, our mission is to cruise what we have.

The waterway begins at Mississippi River mile 598.1, the mouth of the White River. It follows the White for 10 miles to the Arkansas Post Canal, which spans the 9 miles from the White to the Arkansas River. It then follows the Arkansas for 376 miles to its confluence with the Verdigris River, which accounts for the last navigable 50 miles, ending at Catoosa, Oklahoma, which serves as the Port of Tulsa. Along these 445 miles are 17 locks and dams, an unusual concentration for a modern waterway. This is no doubt a partial explanation for the high cost of construction. Another is the extraordinary number of jetties, wing dams, and similar structures needed to control the estimated 100 million tons of silt that descends the waterway each year, constantly striving to outwit the Corps and foil the navigation interests by depositing instant sandbars across channels. The structures accomplish their purpose admirably, but constitute a hazard to recreational boaters accustomed to venturing outside of buoyed 9-foot channels. They are well marked and avoidable if care if exercised; between them can often be found placid harbors with sandy bottoms and beaches.

Pine Bluff, Little Rock, Russellville, Ft. Smith/Van Buren, Muskogee, and Tulsa are the principal urban concentrations along the waterway. Boaters requiring supplies or longing for the bright lights and fast living of the city can go ashore at marinas in the vicinity of the first three and get to town by cab. The Goose Harbor Marina in Van Buren is right in town, within walking distance of all attractions. No access for boaters exists at Muskogee or Tulsa, and the so-called Visitors' Center maintained by the Corps of Engineers adjacent to the lock at Russellville provides no dockage for boaters.

In the vicinity of Little Rock, the scenery changes from Mississippi Delta to Ozark Mountain. Above Russellville, the river becomes a chain of lakes, of which the largest are Lake Dardanelle in Arkansas and Kerr Lake in Oklahoma. The uppermost 50 miles, where the waterway follows a straightened version of the Verdigris River, is the narrowest and arguably the least scenically alluring.

Fuel is available at convenient intervals until you reach the upper end of the waterway. Applegate Cove Marina, at the lower end of Kerr Lake, is the last refueling opportunity. From there to the head of navigation and back is 210 miles, so be sure of your cruising range before you venture much further.

(The author, David F. Ross, is Chairman of the ACS Navigable Canals Committee.

Great flocks of white pelicans make their summer home on the Arkansas.

BOOK REVIEW

Russell Bourne: "Floating West", (New York, N.Y.; W.W. Norton Co., 1992.)

Reviewed by Denver Walton

The story of New York's Erie Canal and its contemporary waterways has seldom been told in a manner as readable as in "Floating West". The author is never boring and sometimes surprising as he details the towpath canal era.

Although the main theme of the book deals with Clinton's ditch and its successors, Mr. Bourne reviews some of the early canal attempts in the U.S., some successful, some not. We learn of Elkanah Watson, who toured Europe's canals, viewing the various means of lifting boats and who was determined to create a passageway westward to the Great Lakes.

Mr. Bourne touches on the plate tectonics that created the various mountain ranges and left us with the handy Mohawk Valley, which enabled the Erie Canal to by-pass the Appalachian barrier.

To prepare the reader for the actual construction of the Erie Canal, the author takes us back to the Revolutionary War days and reviews the Mohawk Valley conflicts with the Iroquois and the British Army.

Several sections in the book provide us with excellent photographs, diagrams and maps, many of them not printed before.

Eventually we came to Clinton, who, more than anyone else, encouraged New York State to build the canal. The antipathy of Martin Van Buren is revealed, but he finally threw his weight behind the canal and the bill was passed in 1817.

We learn much of the building of the canal and what life was like along the waterways, such as the "humiliation" of having to lie down on the roof of the boat while passing a "low bridge" to the unbelievably uncomfortable sleeping arrangements.

Competitive parallel canals were built but none ever made money like the Erie. The Appalachians were the main barrier, and the Portage Railroad, Pennsylvania's solution to the mountain, was a constant problem. One of the most successful canals was the Chesapeake and Ohio, but it never made it across the mountain despite plans for a four-mile tunnel to connect with Pittsburgh.

The canals that remain today, especially the New York State Barge Canal, are largely used by recreational boats.

In conclusion, the author states that "By rediscovering the evocative magic of canals and the mythical people who created them and worked along them, Americans have an opportunity to reclaim a vital connection with the water-land of their birth". This is certainly true in "Floating West", for beginners and experienced canal buffs as well.

"Floating West" costs $24.95 and may be ordered through the publisher, W.W. Norton & Company, Inc., 500 Fifth Avenue, New York, N.Y. 10110. (Shipping costs not known.)

Chicago, Nov. 2, 1992: The Canal Corridor Association announced today that a part of the Water Resources Development Act of 1992 will provide a funding mechanism for capital improvements to the Illinois and Michigan Canal National Heritage Corridor, thanks to Congressman Bill Lipinski.

Canal Corridor Association President Jerry Adelmann said, "For years we have sought to secure federal dollars to match the local commitment to preserving and revitalizing the heritage of our region."

UPPER FOX RIVER AND PORTAGE CANAL HISTORIC SURVEY RESULTS

by Frederica Kleist

On Monday, September 30th, 1991, about 75 people gathered at the community room of the Montello Municipal Building to hear the results of the historic building survey along the upper Fox River system from Portage to Oshkosh. Speaking on behalf of the State Historical Society was Joseph DeRose, historian, while Bill Meindl, preservation planner, spoke for the East Central Wisconsin Regional Planning Commission.

The participants heard the results of the comprehensive year-long survey that identified 1,500 historic structures. Included were houses, stores, factories, resorts, bridges, and the upper Fox lock-dam-canal system. Additional independent surveys will be done in Oshkosh, Berlin, and Portage.

In the slide program, a group from Portage was made aware of the historic downtown of the city. There were several views of the Portage Canal, stressing the importance of this site, which connects the Fox and Wisconsin waterways. A number of Portage residents were involved in the survey.

The Upper Fox River Historical Survey began in 1990, focusing on historic buildings and other structures within a corridor extending 100 yards on both sides of the Fox River and certain adjoining lakes. Areas covered were the Portage Canal, rural properties from Columbia to Winnegago counties, and all or parts of Endeavor, Packwaukee, Montello, Marquette, Princeton, Eureka, Omro, and Butte de Morts.

Results of the survey include:

Identifying 1,500 structures in the survey corridor which are at least 50 years old, and evaluating their history and architecture;

Selecting 275 of these as potentially eligible for listing in the National Register of Historic Places;

Outlining three potential community historic districts—the Main Street Historic District of Omro and separate business and residential districts on and around Water Street in Princeton; and

Identifying 15 lock, dam, and canal properties from Portage to Eureka as potentially eligible for the National Register.

Portage Canal was placed on the National Register August 26, 1977. Other sites eligible are the locks at Montello and Princeton and the locktender's house at Berlin.

This survey produced a comprehensive 35-page historical summary of the Upper Fox River and Portage Canal lock system. The lock system dates back to 1851, and was closed just 100 years later in 1951.

Funding for the survey came from the East Central Wisconsin Regional Planning Commission and the National Park Service, through a grant administered by the State Historical Society. The survey is intended for use by individuals, groups, and communities to help identify, protect, and revitalize their historic buildings, sites, and commercial areas. It can be used to enhance and promote heritage tourism in communities and rural areas along the river and canal.

Persons and organizations interested in the promotion and support of local historic preservation projects can obtain information on placing sites on the National Register from the Wisconsin State Historical Society, 608 264 6500.

(Note: All photos are by Bill Meindl for the East Central Wisconsin Regional Planning Commission.)

The Eureka Lock and locktender's house near Oshkosh are the only lock facilities still in use on the Upper Fox River. The lock here was closed in 1951 but reopened in 1958, and is operated seasonally by the Berlin Boat Club. The lock and house are listed in the National Register of Historic Places.

The Portage Canal was built to connect the Fox River with the Wisconsin River, thus providing through water transportation between the Mississippi River and Lake Michigan. Construction began in 1835.

The Portage Lock on the Portage Canal is a concrete lock built in 1927-28 by the U.S. Army Corps of Engineers. Abandoned 1951.

CUYAHOGA RECREATION AREA

The Towpath Trail from Rockside Road, in Independence, south to Peninsula is "almost" totally open. The new trail has already been used for hiking and biking by many during the year and served as the route for the Toe-to-Tow marathon held on the "Big 160" weekend. Benches and sites for wayside interpretive exhibits are being installed along the trail as are signs and wood posts at road intersections to control vehicle access. A bridge over the Cuyahoga River at the Peninsula Aqueduct site will be installed in the near future.

Lock 38 has been successfully rebuilt, and the gates operate perfectly. Landscaping around the Canal Visitor Center (Locktender's House), including a dike to prevent flooding of the facility, has been completed. Thoughts of "possible" rebuilding Lock 37 at Wilson's Mill are being considered as a future project. The historic Station Road Bridge was rehabilitated and re-dedicated as an important part of the CVNRA trail system.

Boardwalks in Stumpy Basin and at the beaver ponds north of Ira Road are finished. Planning and engineering for next year's work are well under way, and construction of the Towpath is continuing south of Peninsula. The target for completing the entire trail is October, 1993 (The "Big 161", or the 161st year after the *opening* of the entire *Ohio Erie* system from Cleveland tó Portmouth).

The weather, river, and mosquitoes were most unforgiving this year, and the summer rains made towpath rebuilding very difficult in many areas due to continued high water—in some areas the Towpath was inundated. "New" discoveries are still being found, however, and old mills, factories, and a large mule barn foundation have been rediscovered. More treasures are sure to be found, reminding us of our past and showing us a way to a bright future.

Larry Turner

WABASH & ERIE YIELDS WOODEN TREASURE

by Robert F. and Carolyn Schmidt

On June 4, 1991 an historic treasure was unearthed east of Fort Wayne, Indiana on the old Wabash & Erie Canal. A construction crew working on the U.S. 24 acceleration lane from Indiana route 469 accidentally uncovered the remains of a timber-frame lock. The lower third of the lock is in an almost perfect state of preservation, resting in an old glacial lake bed which has kept the timbers packed in moist mud for 150 years. The part remaining is that portion which was covered by the low-water pool in the lock since its construction in the late 1830s.

This wooden lock, known as Gronauer Lock #2, was constructed on the timber-frame design. The dimensions of the lock chamber follow the standard on the Wabash & Erie of 90 feet long by 15 feet wide. Typically for this type of lock, the timbers form cribs which support the interior chamber walls. The cribs adjacent to the gates are stone filled for extra support. Beyond the east gate is an additional 50-foot length of timbered floor and breastwall to protect the canal bed from the water flow rushing from the lock. Further information on this type of construction can be found in the February 1989 issue of *American Canals* in an article by Indiana Canal Society Editor Stan Schmitt.

The level of preservation in the lock is truly remarkable. The timbers are solid, both gate sills are intact, oak planking covers parts of the sides and floor, and the entire lock structure is supported on foundation timbers 37 to 47 feet long. Nails, metal spikes, door strappings, wooden pegs, and portions of the gates have also been found.

Two views of Gronauer Lock #2, unearthed recently by a highway construction crew working on an interchange east of Fort Wayne, Indiana. Some of the old wood timbers are visible in these views.

Lockhouse Located

Across U.S. 24 from the lock, archaeologists have located the basement and foundation of the lockkeeper's house, the cistern, and an artesian well. Substantial information about the lockkeeper, Joseph Gronauer, is known. Pictures of the house and of the Gronauers have been provided by descendants. Also, detailed information on the repairs made on the lock has been found. Archaeologists are impressed by the substantial artifactual remains that have surfaced from the original excavation by the construction crews. The lock was probably used as a trash pit after the canal fell into disuse.

The Gronauer Lock is of historic importance and deserves preservation, not only because of its construction details, but because of its strategic location. It is:

(1) On the longest canal ever built in the United States—458 miles;

(2) Seven miles from where the Wabash & Erie began in Fort Wayne on February 22, 1832;

(3) On the first federal land grant for public works (1827);

(4) Within close proximity to structures in Ohio and Indiana that would link this interstate canal into the historic corridor park plan being considered by the National Park Service;

(5) On the route used in 1846 to take the Miami Indians to the West.

It has definite potential for registration on the National Register of Historic Places.

Based on these considerations, the Canal Society of Indiana has taken immediate action to assure that the unique historic importance of the site is known to all governmental and private authorities at federal, state, and local levels. Letters requesting review of the site have been written to Vice President Quayle, Indiana senators Coats and Lugar, and Governor Bayh. Contact has also been made with our congresswoman and with mayors in Indiana and Ohio. To increase public awareness, society members have been providing daily interpretations for visitors at the site. Thousands of interested people from many states have learned of the importance of this lock. The society has also used the media, presentations at civic organizations, and displays at festivals in both Indiana and Ohio, to build public support. Plans are being made to present the details of this discovery to the 2nd International Conference on Historic Canals at Akron, Ohio, in October.

Currently, all construction near the lock is on hold as archaeologists continue to probe the site. The Indiana Department of Transportation is developing a variety of proposals to resolve this collision between new highways and an old canal. The Indiana Canal Society welcomes any support or creative ideas on preserving this lock for future generations.

Letters in support of the effort of the Canal Society of Indiana to have this site preserved and developed should be directed to: Pat Ralston, Director, Indiana Department of Natural Resources, 612 State Office Building, Indianapolis, Indiana 46204.

Tax-deductible contributions may be sent to: Gronauer Lock Preservation Fund—Canal Society of Indiana, 302 E. Berry St., Ft. Wayne, IN 46802.

A CRUISE ACROSS MICHIGAN

By David F. Ross

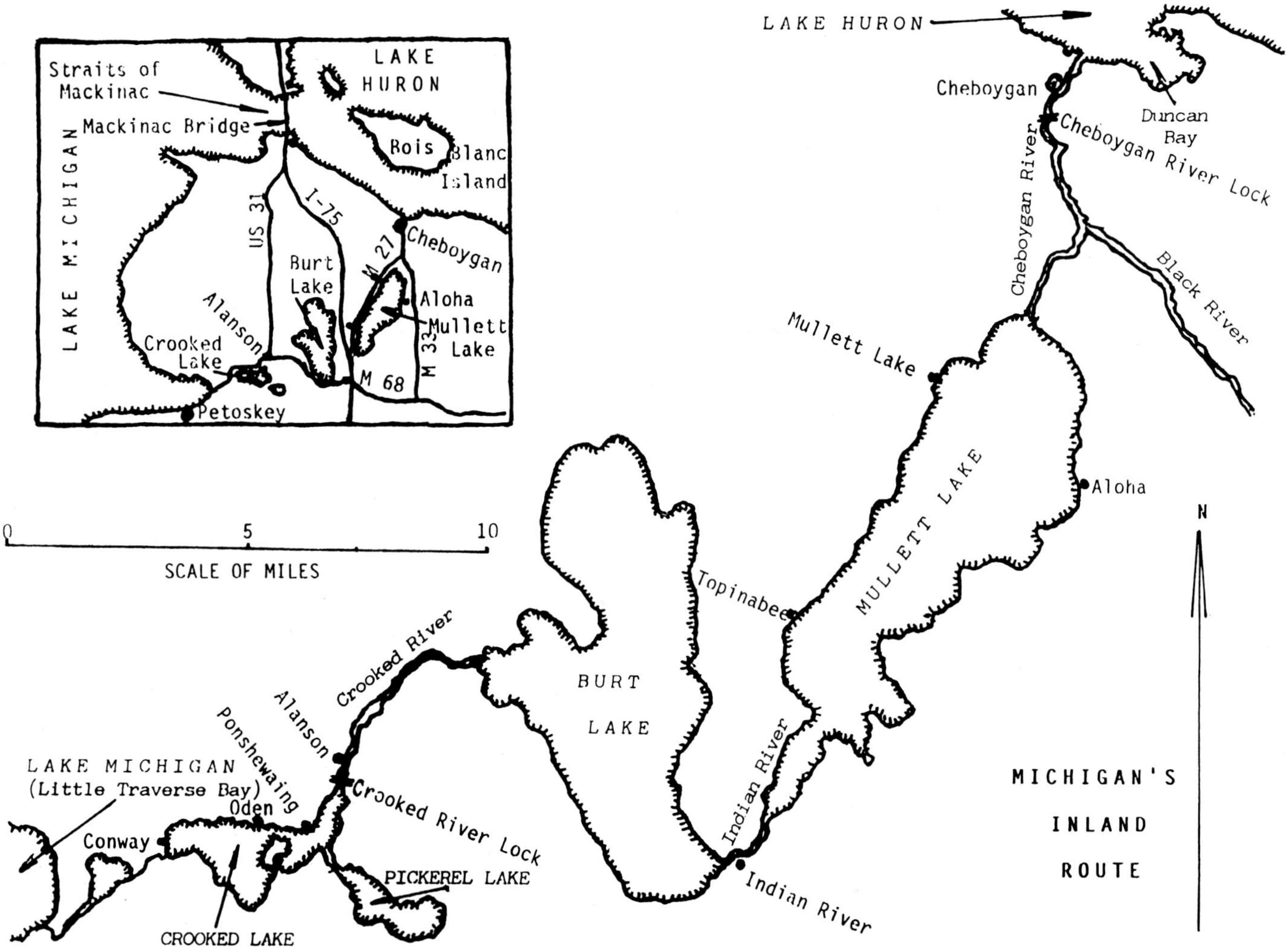

Once upon a time, perhaps about 12,000 years ago, during a period of glacial retreat, what are now lakes Michigan and Huron were connected by a strait considerably to the south of the present Straits of Mackinac. When everything was eventually sorted out, with the ice gone and the water level in equilibrium well below what it had been then, dry land once more predominated where that channel had run. A remnant remained, however, in the form of a chain of lakes and rivers. The highest elevation of the water in this chain is in Crooked Lake, whose western tip is less than two-and-a-half miles from Little Traverse Bay on Michigan's Lake Michigan coast. The outlet from Crooked Lake to Lake Michigan, however, is severely constrained, so that the bulk of the water flows eastward to Lake Huron. Its course is from Crooked Lake via the Crooked River to Burt Lake, thence via the Indian River to Mullett Lake, and finally via the Cheboygan River to the Lake Huron port of Cheboygan. This is a much shorter route from Little Traverse Bay and points south to Cheboygan and points south than the coastal route via the Straits of Mackinac; it is also a much more placid and safer route for small craft, especially canoes. For people traveling by canoe, moreover, the short portage at the western end is only a minor hindrance. As a consequence, this Inland Route (as it is now known) was an important commercial waterway for the Native Americans who inhabited this region before the

The icebreaker Mackinaw in port.

advent of missionaries, soldiers, and settlers of European origin.

The first significant use of the Inland Route by the new inhabitants was related to the 19th century clear-cutting of Michigan's vast pine forests. Sawmills established at Cheboygan and elsewhere could be fed a seemingly limitless supply of raw material at trivial cost by floating logs along the waterway. Before this could happen, however, the logging industry's development required the opening up of the interior to settlement, and the waterway was also the best means to this end. In the 1840s, the Cheboygan River had been perceived primarily as a source of water power for the coastal village of Cheboygan, and it had been dammed. When it came to be seen also as a gate way to the wealth of the interior, the dam became an obstacle as well as an asset. The Cheboygan River Slack Water Navigation Company was formed, and in 1869 it constructed a lock, 18 feet wide and 75 feet long. This is a very small lock for a waterway intended to be used by steamboats, probably limiting access to vessels of less than 50 tons. Perhaps the limitation was dictated by the capacity of the sometimes narrow and tortuous Crooked River. In any case, the Cheboygan River Lock was evidently adequate for the purpose—although operating parts have been replaced as needed, and the gates are now operated electrically rather than manually, the present lock is still basically the original one.

The Cheboygan River Slack Water Navigation Company provided freight, mail, and passenger service to settlements which grew up along the Inland Route from Cheboygan to Conway, at the farthest end of Crooked Lake, and via land connections to Petosky and other communities along the Lake Michigan shore. Its lock also admitted tugs which were used to maneuver the rafts of logs that soon began to fill the waterway, though the logs themselves completed their journey to Cheboygan via a sluice around the dam rather than through the lock. The railroad finally penetrated the northern tip of Michigan's Lower Peninsula in 1881, but, running north and south, it was complementary to rather than competitive with the waterway. Steamboat service continued well into the 1920s, when the automobile and the growing network of public roads made water transport a lark rather than a necessity of life. By this time, the region's economic base was rapidly shifting from lumbering to tourism, and some of the last steamboats in operation were excursion boats based at resort hotels.

The federal government took an interest in the improvements wrought by the navigation company and its successors, especially because of their role in carrying the mail. Financial assistance for channel improvement and maintenance, apparently limited to dredging, was provided. As elsewhere, the gradual conversion of the waterway from commercial to recreational use was accompanied by a decline in federal interest. The difference was that whereas in the typical case the federal government was to a degree wedded to the waterway through its function as the proprietor and operator of the locks and dams, there was no such sacrament of its link to the Inland Route, where the only lock and dam were private artifacts. It took an overt act to break the connection elsewhere, and bureaucratic inertia delayed the inevitable separations until the 1950s or later. On the Inland Route, the government could simply drift away without the trauma of divorce, a much easier process, and accordingly it did so considerably earlier.

Then, in 1954, an astonishing and unprecedented thing occurred. The U.S. Army Corps of Engineers, which was contemporaneously engaged in declaring waterways "obsolete" and abandoning them if they had only recreational use, undertook the restoration and improvement of the Inland Route expressly "to meet the recreational needs of the area." Of course, the corps does not get into waterway projects on its own initiative. Members of Congress from this former lumbering region must have been adept at logrolling. Undoubtedly they were assisted by the expectation that the expense, viewed as a federal budget item, would be insignificant. Little more than dredging was anticipated although jetties were constructed to guard the entrance to the Indian River from Burt Lake. To insure against the possibility that the corps might become involved in the much larger and more binding financial commitments associated with locks and dams, the project was defined as extending from Conway on Crooked Lake to "the upstream end of the lock in the Cheboygan River," even though this meant excluding the last mile-and-a-half of the river as well.

The objective of the project was to establish and maintain a channel with a minimum width of 30 feet and a minimum depth of 5 feet over the 35 miles from Conway to the lock, plus a side channel to Pickerel Lake, which adjoins Crooked Lake. This was evidently a somewhat more generously dimensioned channel than that which had been created in the earlier (19th century) improvement. After the project as originally conceived had been completed, in 1958, it was found that water was being released from Crooked and Pickerel Lakes at too fast a rate, drawing them down to unacceptable levels. To correct this "design error," the corps was obliged to build a weir, and hence a lock, at Alanson, near the head of the Crooked River. This is probably the only lock ever built by the Corps of Engineers expressly for recreational use. It nearly tripled the annual operating and maintenance cost of the waterway (from $28,000 and $80,000, using 1988 data), and has generated chronic controversy and uncertainty over whether the waterway should continue to be maintained, and if so, by whom. The new lock is somewhat smaller than the one at Cheboygan, 16 by 60 feet, and now sets the size limit for boats cruising to Conway. (A still tighter constraint applies to boats going on to Lake Michigan—a marina on Crooked Lake offers portage service, but only for boats up to 30 feet.)

In 1988, the corps undertook a cost-and-benefit study of the Inland Route. These exercises play a major role in waterway development, and people who care about waterways need to understand them. They are, in a nutshell, a means of stating nonsense with precision, and the Inland Route study is an excellent case in point. Benefit data were obtained from a survey of boaters, who were asked what they would be willing to pay for lockage. They were not told what use was to be made of their answers, and many of them naturally assumed that there was to be a lockage fee and that their advice on the amount was being solicited. Everything conspired to produce a downward bias in the responses. Self interest obviously counseled naming a low figure, to save boaters money. On the other hand, a public-spirited boater wishing

(Continued on Page 60)

Cheboygan Crib Light, on Lake Huron at the mouth of the Cheboygan River.

The guillotine gate going up at the Crooked River Lock.

A placid stretch on the Cheboygan River.

A CRUISE ACROSS MICHIGAN

The Cheboygan River lock, open to receive upbound traffic.

(Continued from Page 59)

to show enthusiastic support for the project by stating a high figure would be influenced by the knowledge that the charge for passing through the Cheboygan lock is $3.50. A "high" figure, in that context, might be $5.00, or $7.00, or at the extreme, $10.00. We know that people will pay many times these amounts for the use of ski lifts and golf courses, when the alternative is to forego skiing or golfing, so we may be sure that dedicated boaters would actually pay pretty stiff lockage fees if they had to. They don't really know themselves how high they would go until confronted with the choice in a real-life situation, however, and in any case, the question was not phrased to generate a response in those terms. The result was a finding that the waterway exclusive of the Crooked River Lock generated a benefit-to-cost ratio of 11.4, but that the lock exclusive of the rest of the waterway produced benefits amounting to only 78 percent of its operating and maintenance cost. In their stated terms, these figures are clearly utter and complete nonsense—picture, for example, recreational boaters utilizing a lock while eschewing the waterway of which it is an essential part. What they really mean, however, is not nonsense at all. It is that the corps wished it had never gotten involved with that silly little lock and wanted an excuse to get out of it. At present, both the Cheboygan and the Crooked River locks are being operated by the Michigan Department of Natural Resources, and there is still no user charge at the latter.

If you are trailering your boat to the waterway, there are a great many places where it can be launched. Having tried only three of them, we have settled on Aloha State Park, on the eastern shore of Mullett Lake. The launching ramp is on a protected harbor, and there is no time limit on parking a car and trailer. There is also no charge for either launching or parking, although you must have an official sticker on your car to enter any Michigan state park—about $10 a year, good at all parks. From Aloha it is about four miles north to the start of the Cheboygan River, or about seven miles southwest to the mouth of the Indian River. Mullett Lake is clear and, with depths up to 120 feet, refreshingly cool even at the height of the summer. The entire Cheboygan River, about five miles, is subject to a no-wake speed limit. It is joined at about midpoint by the Black River, navigable for about two-and-a-half miles (farther with good local knowledge of river conditions). A rustic inn, the Hack-ma-tack, provides dockage for diners just after you leave Mullett Lake, as do three other restaurants at the Cheboygan end of the river, two above the lock and the most elegant one below it. There is also a municipal dock below the lock in Cheboygan, from which you can reach all urban services and attractions. Cheboygan harbor is the home port of the 290-foot Coast Guard icebreaker, *Mackinaw,* as well as of ferries to Bois Blanc Island, in Lake Huron. Depending on the weather and the size of your boat, you may or may not want to take yourself out onto Lake Huron. Duncan Bay, just east of the river mouth, offers fairly protected water for those desiring just a taste of Great Lakes cruising. If possible, on a clear day you should at least try to get far enough out on the lake to see the middle of the Mackinaw Bridge, to the west—the ends of it are below the horizon from here.

In the other direction, there are also no-wake zones on the Indian and Crooked Rivers, although not for their entire lengths. You need to enter into the spirit of this, keeping in mind that with a round-trip distance of only about 80 miles, there is no need to hurry, and that a leisurely pace allows you to appreciate the scenery and commune with the swans and ducks. At 30 miles an hour, you might see a great blue heron, but you'd never be able to watch her catch a fish, toss it to get it lined up with her throat, and swallow it. Half way along the Indian River you can particularly appreciate the joys of an unhurried cruise as you pass under the frenetic traffic of I-75. Just west of the bridge is the village of Indian River, with several marinas and a public dock at the foot of the principal business street, in case you want to go ashore. Unlike the much longer one in Cheboygan, however, the Indian River dock is likely to be full

Swans grazing on the Indian River.

The harbor entrance at Aloha State Park, Mullett Lake.

on a summer weekend. The village extends to the shore of Burt Lake.

Burt is similar in size and shape to Mullett Lake, but less than half as deep. The crossing here is about six-and-a-half miles, from the southeast corner to a bay halfway up the west shore. The Crooked River, which empties into the bay, is where you will find most nearly wilderness conditions, although there is one cluster of weekend cottages partway along and, when you are nearly to Crooked Lake, the village of Alanson, Alanson features not only the unique lock already mentioned, but also what it claims is "America's shortest swing bridge." It also offers public dockage, with space nearly always available and all urban attractions within easy walking distance. Soon after Alanson, you enter Crooked Lake, with the village of Conway at its western end. A Conway restaurant provides dockage for diners, and there are two marinas on the north shore of the lake. The channel to Pickerel Lake is on the south shore about a mile from the eastern end. There are no public facilities on Pickerel Lake, but we have found it a more peaceful place to anchor overnight than Crooked Lake, which tends to be fairly rough.

One problem with the Inland Route is its popularity—the effect of a wilderness cruise is somewhat degraded if you meet another boat around every bend in the river. One of our most enjoyable visits was after Labor Day, when the weather was still good and we had the waterway practically to ourselves. A few marinas were closed for the season, but not enough to constitute a problem. At such times, the Cheboygan lock is likely to be unmanned. The locktender, when not at the lock, is at the Department of Natural Resources office nearby. You can go ashore and summon him or her by phone, but more conveniently you can call the Cheboygan harbormaster on channel 16 and ask to have a message relayed. The

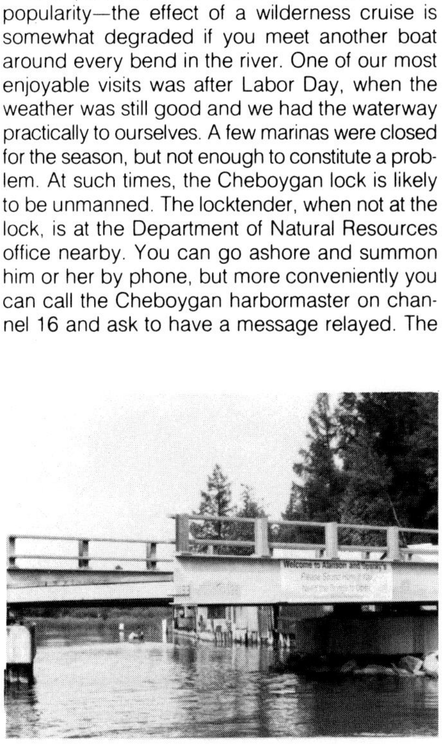

"America's shortest swing bridge," in Alanson.

Inland Route is far enough north so that a September visit will be brightened by autumn foliage. It is also far enough north so that cold weather is possible even in July, so what have you got to lose?

A book of navigation charts for the Inland Route is published as "Chart number 14886" by the National Oceanic and Atmospheric Administration, on sale for $13.25 from the National Ocean Service, Riverdale, Maryland 20737-1199. The route is described in the *United States Coast Pilot*, Great Lakes edition, and *Waterway Guide*, Great Lakes edition. For current information on lockage operations, consult the Michigan Department of Natural Resources, Region II, P.O. Box 128, Roscommon, Michigan 48653, phone 517 275 5151. For information about the portage to Lake Michigan, consult Lake Harbor Marina, Oden, Michigan 49764, phone 616 347 3918.

BOOK REVIEW

ST. CATHERINES: CANADA'S CANAL CITY. John N. JACKSON and Sheila M. WILSON. 414 pp, 320x235mm, profusely illustrated in b & w, maps and diagrams, colour on dust jacket, boards. The St. Catherines Standard Limited. 17 Queen Street, St. Catherines, ONTARIO. L2R 5G5. CANADA. 1992. ISBN 0-919549-24-1. Hard Cover $42.75. Soft Cover $32.05. Mail Orders $2 extra.

The English canal towns of Stourport and Shardlow bear no relation to the Canadian Canal City of St. Catherines. Yet each in their own way started as sites where a local creek offered access to a major waterway. The history of the development of St. Catherines, Canada's Canal City, is carefully detailed in John Jackson and Sheila Wilson's latest book which was sponsored by The St. Catherines Standard to commemorate its centennial. The authors are ideally placed to exploit the local archives as Professor Jackson is Head of the local University Geography Department, whilst Sheila Wilson runs the Local History section of St. Catherines Public Library. These and the paper provided the source for many of the splendid drawings and photographs that make the book a delight for the specialist and the lay reader.

The book is set out in five sections, each of which deal with one era of the town's growth. Waterway enthusiasts will find the section on the founding of the First Welland Canal by William Hamilton Merritt as quite remarkable. He was a man of vision who made things happen. The opening of the canal in 1829 was a triumph indeed. The more so because the water power it brought to St. Catherines provided the basis for a range of industries that provided the backbone to the growth of the City as seen today.

The book details the way in which the pressures to improve the Welland Canal led to the second, third and fourth Welland Canals and how these in turn affected the shape and growth of the town. Surprisingly, little mention is made of the Fifth Welland (By Pass Canal) which has helped the waterway retain its place as a major transport route to the present day.

The life of St. Catherines is set into context as is the way that first the railway and then the Street Cars improved the facilities of the town.

The student of Waterways will find the book most helpful through the selective list of sources arranged by subject that completes this comprehensive book. Others will enjoy it through its readable format which makes the development of the whole area come to life. It is a fitting tribute to Canada's Canal City.

Roger W. Squires

ON THE RICHELIEU AND CHAMBLY CANALS

By Walter Meseck

It was a very worthwhile trip and one that I would recommend to anyone especially those that may have some interest in canals. To me it had particular interest even though I have been over the route several times in power boats. In a way it is strange that we have to go to Canada to get in a canal trip other than Mid-Lakes on the Erie. The Canadians have the Rideau, the Trent and now a Canadian boat is doing their Chambly.

I did enjoy the trip. For the McBride group it is a new operation but I felt they did a good job of pulling it off. Since it was a sleep-and-eat ashore deal (lunch on board) the accommodations ashore ran from good to excellent to "super." The first night we stayed at a good motel right on the river in St. Jean. The second night it was in an excellent motel-hotel right on the St. Lawrence River in Sorel. Dinner and breakfast good and included. At Montreal it was the Queen Elizabeth. What more can you have? All transportation was included and baggage delivered to and picked up to and from your room.

After leaving Burlington we cruised through some of the islands in the upper part of Lake Champlain before entering the Richelieu River for a stop at Fort Lennox, a fort the Canadians built to keep those damned Yankees where they belong. We built one too but we made the mistake of building it in Canadian territory, "Fort Nonsense."

A good part of the afternoon is spent going down the Richelieu River through farm country, summer cottages and some really nice homes. Overnight at St. Jean. In the morning we go through the first lock at St. Jean and for most of the morning and early afternoon are going through the series of locks that will put us back in the Richelieu River at Chambly, thus by-passing the rapids in the river. Below Chambly it was a series of small towns, farms and homes with another overnight stop at Sorel. In the morning it was up the St. Lawrence River to Montreal. This part was of interest to me because of the commercial traffic. There was not as much as I would have liked, so things must be slow in Canada also.

CANAL STAMPS

Colorful British Canal Stamps are now available in connection with the Brits' celebration of "CANALS 200". Write the British Philatelic Bureau, 20 Brandon St., Edinburgh, EH3 5TT, United Kingdom, for information.

MISSISSIPPI RIVER MUSEUM THREATENED

by John W. Spence

A long-running soap opera in Memphis-on-the-Mississippi has been "Sidney and the Pyramid," a tale of a smooth-talking promoter brought to the West Tennessee city by well-meaning financiers to make money for himself and the city and county governments by cramming "The Great American Pyramid" with tourist-attracting entertainments. Closely linked in the public's view yet under a separate contract is the management of Mud Island and the Mississippi River Museum.

As this was being written, promoter Sidney Shlenker and his partner, John Burton Tigrett, faced a deadline to substitute $3 million in cash for common stocks in an escrow fund intended to protect Memphis and Shelby County in the event of nonperformance by the Pyramid Company on the 25-year management contract for the Pyramid. Memphis Mayor Richard Hackett and Shelby County Mayor William Morris has gone to court asking Shlenker & Co. to put up the money by May 16.

Shlenker was introduced to Memphis by Tigrett, a 78-year-old Memphian thought to be worth many millions himself. Shlenker was said to have made a success in Denver with a professional basketball team and to have managed Houston Astrodome successfully for that city's famed promoter, Judge Roy Hofheinz.

It is a tangled story for at least three reasons. First, the construction of The Great American Pyramid, basically a sports arena to seat 20,000 or so at indoor sports events, is under a separate contract which is going well. Second, there is Shlenker's management contract involving securing concessioners, other attractions, and the fittings required for these. Third, and most important in the minds of the minds of many persons in Memphis and the Mississippi Valley, there is the Shlenker-Hackett 20-year contract, signed in the spring of 1989, giving Shlenker virtually unlimited freedom to alter the already-existing complex of attractions on Mud Island.

The focus of this article is on Mud Island and the Mississippi River Museum. These two are indeed inseparable physically. Separate managements for Mud Island's many and varied facilities existed from 1982 to 1989. Mayor Hackett, independently of Memphis' 13-member City Council, decided the city's taxpayers should be relieved of the expenses incurred each year in opening Mud Island to the public. It is probable that, had Hackett asked for council approval at that time, he would have secured it, making moot an issue that arose later, the issue of a Memphis mayor's authority to sign contracts without City Council approval.

Mud Island as the world now knows it was the dream of a highly regarded architect who had designed such successful Memphis landmarks as the International Airport, the Memphis College of Art, the city's most successful department store, public schools, and other elegant buildings. This is Roy Harrover.

From his office windows in downtown Memphis, Harrover, like thousands of other Memphis business people and apartment residents, could see Mud Island. The island was an unsought gift to the city from the silt-laden waters of the continent's most important waterway, the 2,500-mile-long Mississippi River.

Present on Memphis' doorstep since 1913, the island was always subject to overflow when the Mississippi flooded. Short-season crops and squatters were the main occupants of the mile-long island of some hundreds of acres. After World War II, the island was graced with a modern-day use: a very popular one-runway general-aviation airport. It's major drawback: access, other than by air, was by boat only. No automotive bridge linked Mud Island to the mainland until about six years ago. Wolf River separated Mud Island from the Memphis riverfront, flowing in a southerly course on the east side of the island to join the Mississippi about the west end of Beale Street. About 30 years ago, the city and the Corps of Engineers agreed to divert the Wolf River straight west across the north end of Mud Island, damming the north end of the Wolf's old channel and creating a roadway link on the dam to the island. But this access was so distant, so inconvenient for downtown business people, that use of Mud Island continued to languish, until Harrover had his vision. What he conceived was a unique riverfront, river-oriented park and museum, an ensemble of music, dining, children's and adults' entertainment, and—most important to Harrover and his supporters—the Mississippi River Museum, a unique portrayal of the Mississippi River's history and economic use from prehistoric times to the present day.

Modern museums frequently offer a "hands-on experience." At Mud Island's Riverwalk, however, it is a feet-on experience.

Multiple Use Concept

This multiple-use concept has been both the making and the breaking of the Mud Island plan. Harrover overcame the lack of linkage with a magnificent, and costly, monorail passenger conveyor topped by a quarter-mile-long pedestrian walkway. The western terminus of the monorail is the Mississippi River Museum. From that point, ignoring the museum if they chose, Mud Island visitors could walk to shops, restaurants, and a 5,000-seat amphitheater. They could also enjoy the "Riverwalk," a 1,400-foot-long replica of the Mississippi River, accurate in detail, from the junction of the Ohio and the Mississippi to the Gulf of Mexico. Regions west and north of Cairo, sources of the Mississippi, were also carefully depicted in vertical images. Tens of thousands of island visitors have tucked toes into this surrogate Mississippi, or have found and stood upon the miniature sites of their own homes in the cities along the river: Cairo, Carruthersville, Oseola, Memphis, Helena, Greenville, Vicksburg, Natchez, Baton Rouge, New Orleans. Harrover intended the Riverwalk to climax visits to the Mississippi River Museum. "It is an integral part of the museum," Harrover says today. "It's one of the most educational features of the museum as well as one of the most enjoyable features. Managers of the museum should manage the Riverwalk."

Originally, the city set up an independent board to operate Mud Island, its Mississippi River Museum, the shops and restaurants, the marina and amphitheater, and the children's playground. Here began the difficulties. Commercial operators contracted to run the shops and restaurants and to bring entertainment to the amphitheater. The overall operation did not meet both current expenses and capital amortization charges. About 1984, Mayor Hacket asked the City Council to turn management over to a long-established and respected unit of city government, the Memphis Park Commission. The Park Commission at that time was less experienced in museum management than it has since become. At that time its major museums were a natural history museum called The Pink Palace, an antebellum cottage, and the Lichterman Nature Center. To its credit, the commission saw its needs and brought about the creation of a semi-independent Memphis Museums Corporation, delegated responsibility, and today oversees an expertly managed museums system.

Alas, it was too late for the Mississippi River Museum to benefit. Separate records of expense for Mud Island's many facets were not kept. Utilities, security, maintenance, staffing, and insurance were not, as far as the public has ever known, prorated among the shops, restaurants, amphitheater, and museum which used them jointly. Attendance records for the museum were kept separately—admission is charged for the monorail and again for entrance to the museum—but that is the only available clue to the popularity of the river museum. It had been popular.

63-Million Investment

What Mayor Hackett saw, however, was a $63 million investment by taxpayers that was costing more money each year. Beyond anyone's control in the period of Mud Island construction—the mid-70s to 1982—was the nation's worst period of inflation. Construction costs far exceeded first estimates, disgruntling the City Council and the Mayor, and inflating the costs of amortizing bond principal and interest. Mayor Hackett saw in Shlenker a man who could "turn a profit" for the city and himself as boss of the whole of Mud Island, museum included. Some members of the City Council took issue with Hackett's belief that the city charter authorized him—without council approval—to sign away control of a vast city property, Mud Island, for 20 years. The issue was not put to a judicial test, but a referendum on the issue was on the ballot in the fall of 1989, and the voters sided with Mayor Hackett.

Unforeseen in April 1989, when Hackett and Shlenker signed the 20-year deal, was a downturn in the U.S. economy. That trend, coupled with—

The steamboat era receives due attention at the Mississippi River Museum.

The Riverwalk, a scale model of the Lower Mississippi, is one of the most educational and entertaining features of the museum complex on Mud Island.

some say—egregious errors in Shlenker's part, have meant that the sometime Denver basketball magnate failed again and again to borrow the money he needed to make good on his two contracts to develop The Great American Pyramid and to convert Mud Island to a profit-making operation. The Pyramid has been much more the focus of attention for the mayors and the city's only major daily newspaper, the Commercial Appeal.

But it has been Shlenker's seemingly offhand announcements of his plans to break up the museum and disperse its contents among a series of bars, nightclubs, and restaurants which evoked sharp negative responses from ordinary Memphians and two organizations. Private citizens voiced their outrage in calls to Mayor Hackett and letters to the editor. The two organizations, the Mud Island Foundation and the West Tennessee Historical Society, entered the battle to save the museum in the summer of 1990.

Mud Island Foundation

The Mud Island Foundation is a small group of civic-minded Memphians who secured a State of Tennessee charter to support a not-for-profit operation of Mud Island. The group came into being just in time to publish a handsome multi-color "Complete Guide to Mud Island"—and to see its plans of promoting the island and the museum dashed by Hackett's transfer of them to a for-profit operator.

The West Tennessee Historical Society, long established and with hundreds of members, went first to Mayor Hackett and later to Shlenker for assurances that no permanent and irreversible changes would be made in the Mississippi River Museum. Hackett assured all that, at least for one year, no structural changes would be made. Hackett also voiced his disapproval of Shlenker's ideas for mixing bars and museum artifacts.

But a one-year grace period for a multimillion-dollar, unique museum has seemed too little to the Mud Island Foundation and the Historical Society. Could or would Mayor Hackett renegotiate the contract with Shlenker to return the museum to not-for-profit operation under a quasi-public organization or a joint public-private arrangement?

Hackett's preoccupation with myriad other city problems—from call-ups of city employees in military reserves for Gulf War duty to the successful opening of an exhibit of Russian art (the "Catherine the Great Exhibit") has left this question unanswered.

The foundation and the historical society sought to help answer the question by bringing to Memphis last fall experts in museum design and management from the Zenda Inc. firm in Nashville. Zenda's principals, Charles Phillips and Patricia Hogan, concluded that "[A]lterations currently planned . . . would . . . destroy the integrity of . . . the Mississippi River Museum . . . [and] would also have a detrimental effect on the cultural vitality of Memphis and cause difficulties for the city's other museums." They also pointed out that "The Mississippi River Museum was . . . built around a specific interpretation of the Mississippi River and its environs. The very construction of the building centered around the interpretive plan. This is one of the reasons that the Tennessee Historical Commission's recent resolution was so insistent on keeping the museum as it is in the building it currently occupies."

Photographs by Murray Riss for the Mud Island Foundation, used by permission.

ANOTHER ERIE CANAL GUIDE PUBLISHED

Cruising the Erie Canal and South: A Complete Guide to the Canal, is the title of a recent work by Tom Kranz. It is described as covering the Niagara River, Lake Champlain, the Hudson River, and the canals connected to them. The book may be purchased from Longacres North American Entertainment, 1529 Mill Rd., East Aurora, NY 14052. $14.94 plus $2 for shipping plus 8 percent N.Y. sales tax where applicable.

GRONAUER LOCK UPDATE

In August 1991, we reported on the unearthing of the Wabash and Erie Canal's Gronauer Lock near Fort Wayne, in the course of construction of a highway interchange ("Wabash & Erie Yields Wooden Treasure," by Robert F. and Carolyn Schmidt). Everyone now seems agreed that the lock should be preserved, but disagreement is intense over such questions as where and at whose expense the preservation should be effected. The Indiana Department of Transportation favors removal of the lock to a site accessible to the public but not directly in the path of a public road. The department has offered to fund the removal, to the extent of $100,000, if someone else can be found to take responsibility for the lock thereafter. The Canal Society of Indiana is lobbying vigorously for relocation of the interchange and preservation of the lock in situ. Meanwhile, whatever the outcome, the discovery and the resultant controversy are reported to have produced a healthy awakening of interest among Indianians in their canal heritage, and a healthy increase in membership in the Canal Society of Indiana.

MEMPHIS TO NEW ORLEANS ON THE MISSISSIPPI QUEEN

When we returned from our taur of Natchez, Mississippi we found the "Delta Queen", sister steamboat to the "Mississippi Queen", tied up alongside us. They were headed north, while we were headed south.

The "Delta Queen" as viewed from the upper deck of the "Mississippi Queen" at Natchez.

By Bill Shank

Thanks to my sister, Betty Post of Houston, I realized one of my long-time ambitions—to ride the "Mississippi Queen" on a week-long voyage (April 3-10, 1992) along the lower Mississippi.

I traveled with a group of some twenty people, mostly retired, from the Memorial Drive Presbyterian Church of Houston, Texas, led by the Rev. Harry Adamson, all of whom I met in Memphis. Unfortunately, my sister developed a severe arm infection at the last minute and had to stay home.

Our trip on the "Queen" included stops at Vicksburg, Natchez, St. Francisville, Baton-Rouge, and finally—New Orleans. We were favored with excellent weather for the entire trip. The "M.Q..", built in 1975, is one of two steamboats operated by the company, the other being the "Delta-Queen" built in 1927. The M.S. is the larger of the two and is the biggest all-steam-powered luxury vessel in the world. She carries 416 passengers, plus crew of 156.

The service on board is comparable to the finest hotels in the country. The entire ship's crew, officer personnel, room and table attendants make you feel at home from the moment you step on board until you disembark at your destination.

(Concluded on Page 65)

C & O CANAL BOAT TRIPS

The boating season has resumed at the Chesapeake & Ohio Canal National Historic Park. A.C.S. members visiting the nation's capital will want to take time from their hectic schedules of lobbying, legislating, log-rolling, and legal-loophole-locating to slip back in time and enjoy a peaceful and instructive cruise on one of the Park Service canal boats. Both are crewed by Park Service personnel in authentic period costumes and pulled by authentic Park Service mules in their natural state.

The **Georgetown** operates from the Georgetown Visitor's Center at Foundry Mall, 1055 Thomas Jefferson St., in the District. The **Canal Clipper** is berthed at the C&O Museum, Great Falls Tavern, 11710 MacArthur Blvd., Potomac, Maryland, just outside the District. Both vessels schedule 10:30 a.m. and 1:00 and 3:00 p.m. departures Wednesday through Sunday. There are no Monday and Tuesday trips.

Trips include passing through a lock, canal songs, and instruction in the history and legends of the canal. For further information, call 202 653 5844 or 301 443 0024.

[Information by the C&O Canal Association.]

A Union "Iron-Clad" Warship, sunk by Confederate canons at Vicksburg, and recently raised from the mud of the river bottom. Now on display at the Vicksburg Battlefield.

"End of the line" for the St. Charles Street Trolley Cars, on the outskirts of New Orleans. Several dozen of these cars have been running constantly since the early 1900's, bringing commuters into the city.

ON THE MISSISSIPPI QUEEN

The Captain's Dinner Party guests pose on the elegant stairway leading to the grand saloon and restaurant. Front row, left are Captain Keeton and his wife. Directly behind Mrs. Keeton is Bill Shank, who was her dinner-companion.

At each major shore stop we were offered bus tours, which were well organized and educational. I had no idea of the importance of the Battle and seige of Vicksburg, where the Confederate garrison held out for 47 days against the onslaught of Yankee forces, from both upstream and down.

The Vicksburg battlefield is almost as large and well documented as the Gettysburg battlefield. The day Vicksburg surrendered was the same day as the disastrous Pickett's Charge at Gettysburg! Below the battlefield we found the well-preserved remnants of a Yankee "Ironclad" which had been blasted and sunk in the river by the Confederate guns and dug up again just a few years ago.

Natchez is a beautiful city with many ante-bellum homes there in the style of "Gone With the Wind" mansions. At Baton-Rouge our memory of the colorful careers of the late Huey Long and his brother, Earl Long, were brought back to life for us by well-informed tour guides. The State Capitol is an imposing building, with a statue of "The Kingfish" dominating its near-by gardens.

At New Orleans, I ignored the new and elegant trolley-car running along the harbour-front in favor of the nearly century-old St. Charles Street trolley car line, which, with its many operational cars, looks, feels, sounds and rides like the many old cars I remember riding in Pennsylvania as a young boy. Talk about nostalgia—you have it here in large doses; I rode to the end of the line, and it only costs a dollar one way!

I think I made a lot of friends for the American Canal Society on board the M.Q. I was even invited to the Captain's reception and dinner one of the last nights on board, as was also Harry Adamson. I left much canal propaganda with the executive officer for the ship's library. I recommend this trip highly for any canal or river buffs who want to see any part of the Mississippi and its tributaries, from the deck of a steamboat. Write Delta Queen Steamboat Company, Robin Street Wharf, New Orleans, LA 70130. Allow plenty of "lead time"; the Captain's wife told me most of their trips are sold out two year's ahead.!

(Concluded from Page 64)

The huge paddlewheel (36-feet wide by 22-feet in diameter) at the stern of the boat provides adequate power, upstream and down, to drive the 382-foot long by 68-foot wide steamboat, and into and out of each port-of-call along the river. At Baton-Rouge we tied up at a floating dock, with gangplank, but elsewhere we used the ship's own gangplank suspended from a crane at the bow.

Mud Island Museum at Memphis, from the deck of the "Mississippi Queen". (See Issue No. 77 of A.C.)

As we left every port-of-call, day or night, the Caliope on the stern of the "Mississippi Queen" tuned up for a concert to the local citizens. The operator can be seen at the console of the steam organ in the lower right foreground.

A view of the "Mississippi Queen" tied up at Vicksburg, Mississippi taken by your correspondent (stern view). Some idea of the tremendous size of this ship can be gained from this view. The vessel draws only eight feet of water, ideal for the sometime shallow channel of the river.

"SPRUCE GOOSE" TRAVELS THROUGH WILLAMETTE FALLS LOCKS

The "Spruce Goose" enters the Willamette Falls Locks. Photo by Bill Johnson

Oct. 15, 1991, the U.S. Army Corps of Engineers' Willamette Falls Locks became history. They were formally designated as a State Historic Civil Engineering Landmark by the History and Heritage Committee of the American Society of Civil Engineers. Constructed in 1870 and opened on New Years Day in 1873, the Locks are the oldest locks in the West in operation today.

Dec. 17, 1992, history happened again, as the world's largest airplane, Howard Hughes' Hercules HK-1 Flying Boat, traveled through the Portland District Willamette Falls Locks, on its way to McMinnville, Ore.

"When the locks were built in the 1870s, who would have thought they would be moving the body of the world's largest airplane through it?" said one Corps employee as he watched the flying boat, better known as the "Spruce Goose" approach the lock chamber.

John Wasson, Portland District's project manager at Willamette Falls Locks, however, had been preparing for the journey for quite some time, measuring the locks time and time again to assure a safe passage. Despite careful calculations, one tiny glitch developed—the river was lower than ideal.

"We're doing it a little different—with one of the lock gates open," Wasson said, as he made final preparations for the wooden Spruce Goose to travel through.

Normally, a vessel will lock through each of the five locks one at a time. When in the first lock, miter gates on the downstream end of the lock chamber close to hold incoming water, raising the vessel to the next lock. Then, upstream miter gates open, allowing the vessel to travel to the next lock, and so on.

On this occasion, however, the water level was too low to allow the Goose's nose to pass over the top of the second gate. Both the first and second lock chambers had to be opened, leading to another possible glitch. Rock, shale, logs and bark lay at the bottom of the chamber where smooth concrete surfaces rest in modern locks.

Made from native stonework ranging from five to 15 feet, which remain watertight after 120 years, the locks reflect construction methods before concrete technology was developed.

The challenge to filling both chambers at the same time was keeping the "bottom" of the second lock from travelling with the incoming water to the first lock and lodging underneath the barge or against the 3-foot sill elevation of the second lock gate. Only one other time had two lock gates been open at the same time.

"If we let the water out and wash all that in, it could be a fiasco," Wasson said. With the same breath, Wasson added, "It's all going to pull together," encouraging himself and those nearby. "They have the most professional people [towing and barging it]," he said. "They" were Tidewater Barge Lines from Vancouver, Wash., "the biggest tow boating company on the Columbia," and Joe Bernert Towing from Wilsonville, Ore., he said, who were responsible for river travel of the flying boat.

Wasson's encouragement was right on. The "Spruce Goose," completed in 1947, finally approached the lock chambers, the "bottom" stayed put and onlookers seemed awed at the immense sight.

Pushed by "Rainbow B.," a green and white tug boat belonging to Joe Bernert Towing, the "Spruce Goose" floated regally above everything. The water slowly rose so that the barge and its cargo could pass through to the third lock chamber, where the water would rise high enough to place the barge at almost eye level, hoisting the goose far into the air.

The land travel, coordinated by Emmert International, will end when the "Spruce Goose" is delivered to the Evergreen AirVenture Museum, where it will stay until its own museum is constructed.

When it is fully assembled in the museum, its appearance will continue to be mind staggering, Gifford said, who saw the plane when it was assembled at Long Beach, Calif.

"You see those C-141s and think 'those are enormous,' then you see something like this that was invented 40 years ago . . . it dwarfed everything. I've never seen anything as large that had a service to it," he said.

Despite Mother Nature's attempt to botch the enjoyment with rain, wind and cold temperatures, the idea that history was being made in a very big way overrode the elements for many.

"It's a once in a lifetime . . . we'll never see a plane go through like this again," said Tom Currany of Dundee, Ore.

And he's probably right. The longest trip by the world's largest airplane through the oldest wooden locks will probably only happen once. And, after 120 years of operation, for the Willamette Falls Locks, that's all it takes.

BOOK REVIEW

Through a blend of evocative prose and museum-quality photographs, Author Jim Redd entices us to join him on a personal journey down the Illinois and Michigan Canal.

In *The Illinois and Michigan Canal: A Contemporary Perspective in Essays and Photographs* (136 pages, $29.95, May 12), Redd shares the experience of discovering the canal's archaeological record, of learning its history, and of meeting the people who live along its corridor. He celebrates the canal's aqueducts, locks, and masonry as representative of an earlier, more humane form of technology and engineering.

The Illinois and Michigan Canal runs from Des Plaines to La Salle. Completed in 1848 and closed to navigation in 1933, it symbolizes the final flourish of preindustrial technology. Redd finds the canal fascinating "partly because of its historical significance but also because of the inherent visual richness I have found in its old masonry structures and the towns it links across a hundred miles of Illinois landscape."

Yet reality tempers this nostalgic view. Redd's camera finds the "rusting ruins of twentieth-century industry." And he writes of the "stench of stagnant water and industrial waste." To Redd, however, the "authenticity of this unmoderated harshness was engaging in its own way, and I have made no effort to disguise this present reality in the photographs."

Redd balances the objective documentation of the photographs against his subjective account, which germinates in history and blossoms into an intimate sense of what the canal has been and is now. He sees beauty where most of us would see only ruins; and, in this saga of the land as a record of everything that has happened throughout time, he shows us what we missed.

Obsolete now, this is the canal that catapulted Chicago from village to major metropolis.

LOUISIANA'S B&L CANAL

By William Dzombak

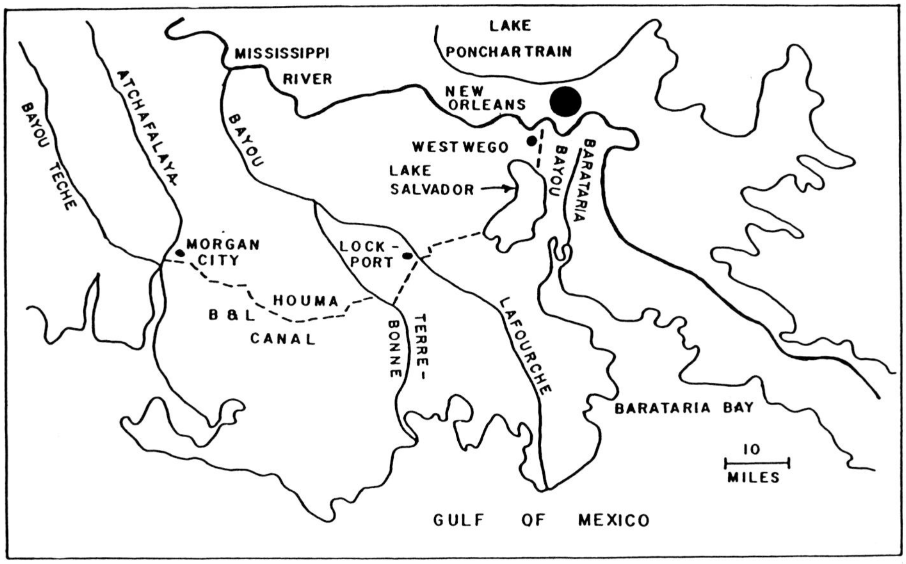

A recent book [1], reviewed here, details the three-part history of an east-west canal, across the Mississippi delta, that was begun in 1829 and eventually became part of the Gulf Intracoastal Waterway, started in 1925 by the U.S. Army Corps of Engineers.

By 1830, many large cotton and sugar cane plantations had been established in southern Louisiana, west of New Orleans, on the rich alluvial soil deposited by Mississippi River floods, but transportation of crops to market—to New Orleans, that is—was restricted to limited navigation in a north or south direction on the major distributaries of the Mississippi River, particularly these: the Bayou Lafourche, Bayou Terrebonne and, further west, the Atchafalaya River and the Bayou Teche. There was no direct route eastward, to New Orleans, across the swamps and marshes that lay between adjacent north-to-south bayours. The only alternative was to boat down a bayou to the Gulf, then boat up an adjacent bayou to reach a neighbor or settlement situated only a few miles east or west from the point of departure. A flatboat journey onto the Gulf of Mexico was too risky to attempt.

The first east-west canals on the delta were mere ditches, or trails, scooped out of the swamp muck and just wide enough and deep enough for small boats. The first navigation canal was dug in 1740, by Claude Dubreuil, to connect New Orleans directly southward to the Gulf via the Bayou Barataria. This canal brought building materials to New Orleans—cypress timbers and clam shells (lime) found on middens left by Indians. Before the Battle of New Orleans, in 1815, pirate Jean Laffitte used this canal to smuggle prizes into and out of New Orleans by avoiding the guarded Mississippi River route to the city. Dubreuil's canal later was known as the Gardere Canal.

Parallel to the Gardere Canal, Nicholas Destrehan dug another canal southward to the Bayou Barataria. Destrehan's son-in-law, Joseph Harvey, developed the channel as the Harvey Canal that eventually was to become part of the Intracoastal Waterway, in 1925.

The only way to get to the plantations in the west, near the Bayou Teche, was to go 50 miles or more up the Mississippi River above New Orleans, then to venture onto one of the bayous—Plaquemine, for example—that headed westward, more or less. That roundabout way was not only long, but also dangerous because it could be used only at times of flood and never could accommodate steam boats. At most other times of the year, the water was too low for navigation, and was made hazardous by submerged logs.

The Erie Canal was opened, and began to return large revenues, at a time when sugar plantations were being created on the Louisiana delta. Success on the Erie must have spread canal fever to Louisiana because the much needed west-to-east navigation canal, called the "Barataria and Lafourche Canal," was started in 1829 by Dr. Walter Brashear, who operated a plantation in the Teche country, and Judge Charles Derbigny, of New Orleans, who was the son of the Louisiana governor. Their plan was to dig short east-east canals to link natural waterways—major bayous—that flowed in the right direction, more or less. A major part of the route they proposed was to follow the Bayou Black, which ran westward from Houma. Locks would be needed only where the east-west canal crossed the major bayous, which generally flow from north to south and empty into the Gulf. Periodic floods on the bayous filled the canal segments with deposits of mud and debris; dredging, and clearing snags, proved to be an endless maintenance headache.

Building locks founded on river sediments posed problems. The marshy terrain, called "trembling prairies" by the early settlers, thwarted railroad construction and so explains how it was possible for the B&L Canal to continue to operate profitably into the 20th century. The first locks built were made of cypress timbers—at Lockport, for example, where the B&L Canal crossed Bayou Lafourche. At each such crossing, a set of two locks was needed, one lock on each side of the stream. Seasonal floods scoured away the foundations of the locks. Decades later, the state assisted with the construction of more substantial locks built with bricks and hydraulic cement. On those masonry locks, the gate sills were cypress, but the gate post quoins were cast iron. At the Mississippi River end of the canal, where Westwago now stands, the lock was to be 25 feet wide and the gates 130 feet apart; at low water, there was to be at least 6 feet of depth over the gate sills; the locks were made high enough to be at least 2 feet above the high water level in 1849.

In 1852, when the masonry locks at Lockport were being built, it was necessary to finish the high walls and gates on the Bayou Lafourche while the water level was low. To assist the ninety-six slaves who were working on the job, extra laborers were hired—white men, but it was found that they could not do half the work done by the same number of Negroes. Occasional floods caused suspension of the work. Mismanagement compounded the difficulties associated with marshy terrain, floods, sedimentation, and fluctuating water levels.

Taxpayers objected to state subsidies for a private enterprise that charged tolls on natural waterways, and the state engineers realized that building and maintaining waterways on the delta was an enormous, never-ending task that perhaps required the resources of the United States Government. In 1859, Louisiana withdrew its support for the B&L Canal project and transferred its stock and debt responsibility to a new company that would be formed. After 30 years of effort, the original Barataria and Lafourche Canal Company had failed to complete the canal. The bankrupt "Company Canal", as it was called, was picked up by Robert Barrow, a wealthy planter who owned several plantations and worked 700 slaves.

Barrow took over the canal project in 1859 and formed the Barataria and Lafourche Canal Company No. 2. His intention to complete the canal was interrupted by the Civil War and the disruption of plantation work and production. The slaves had departed. After the war, Barrow tried various schemes and ploys designed to further the canal project, but his efforts did not succeed. When he died in 1875, the B&L Canal still had not been completed, although some sections had been navigable for a time.

Robert Barrow, Jr., at age 17, took over the B&L Canal project that his father had tried to complete. At the same time, the Corps of Engineers began a program of clearing and dredging all navigable streams, including Bayou Lafourche, Terrebonne, and Bayou Black. Barrow's son sought to capitalize on that indirect federal assistance. He would restore the B&L Canal links to the bayou channels dredged by the Corps. He succeeded. In 1911, he was able to say that the entire length of the B&L Canal was in operation, from Westwago, on the Mississippi River opposite New Orleans, as far west as the Atchafalaya River.

By then, railroad spurs had been built to run southward from the northern edge of the swampy region. Those spurs siphoned freight away from the B&L Canal. The future of the canal looked bleak. Added to the usual maintenance problems was the growth of water hyacinth plants that spread throughout the region and choked the canals. Robert Barrow, Jr. had long been looking for someone who would buy the B&L Canal. The federal government became that buyer, in 1925, when the Corps incorporated the B&L Canal into the Gulf Intracoastal Waterway.

Thomas Becnel's book on the Barrow family and the Bayou Barataria-to-Bayou Lafourche canal is replete with details pertaining to the political and financial machinations of the Barrows, as they pursued a canal vision that had begun with short ditches scratched for Frenchmen in pirogues. There is not much in this book about canal engineering because B&L was essentially a water-level route from end to end. Maintenance was the main concern, dealing with floods, flotsam, snags, mud slides and sedimentation. Engineering was limited to the problem of building locks on wet soil, without bedrock. The story of the B&L is interesting because that canal was built on "difficult" terrain, where water supply was a problem of a different kind.

[1] Becnel, Thomas A.
The Barrow Family and the Barataria and Lafourche Canal (The Transportation Revolution in Louisiana, 1829-1925) Louisiana State University Press, Baton Rouge, 1989.

CRUISING THE CANADIAN CANALS

Robert Akers poses for his picutre at the bow of the "Kawartha Voyageur." Note the pilot house above the second deck.

By Robert Akers

The canal systems of Canada offer idyllic scenes, lots of locks, friendly people and a chance to see 1892 living in 1992 comfort. My wife and I took the Quinte Summer cruise offered by Ontario Waterway Cruises in September.

The company operates three five day segments on canals, rivers and bays along Lake Ontario. We took the center portion which started at Kingston on Lake Ontario and ended at Peterborough on the Trent-Severn Waterway. It was five informal days of excellent meals, delightful harbors and ever-changing views through 19 locks. The passengers were well-traveled and quite entertaining.

Our boat was the Kawartha Voyageur operated by the Lloyd Ackert family. The second generation, Marc and John handled piloting, maintenance, social activities, commentary and general management. Their land support, supply and booking arrangements are handled by their respective spouses, Heather and Joy. The crew consisted of Sharon, the cook (and a good one she was), deckhands Kelly and Barb, and servers Helena and Peggy.

John's son Ryan joined us for three days of the trip and then much to his distress he had to return to school. At age nine, he is determined to follow his father and grandfather and work in the family business.

The boat accommodates 24 passengers in 12 cabins on the lower deck. The cabins are snug but adequate with passengers responsible for their tidyness during the voyage. While usually full, our trip carried just 13. The lounge and dining room are on the upper deck with a sun deck above for viewing the birds, flowers, resort homes and other boats passing by.

Our voyage started at Kingston on Lake Ontario with scheduled overnight stops at Picton, Trenton, Percy Reach and Hastings. Normally the first night is spent at Kingston, but Marc elected to leave at 8 p.m. to avoid an anticipated storm in Lake Ontario. We sailed in the darkness until about midnight which was unusual and enjoyable and tied up at Prinyer's cove. We cruised the Bay of Quinte to Trenton and then traveled on the Trent River passing through 19 locks en route to Peterborough.

It was a pleasant voyage with pleasant passengers and a competent and friendly crew. We are looking forward to taking other segments offered from Peterborough to Big Chute off of Georgian Bay and from Kingston to Ottawa on the Rideau Canal.

The Trent-Severn Waterway and the Rideau Canal are operated and maintained by the Park Service of Canada. The canals and dams were built originally in the 1800's to insure adequate water for log drives and milling. They are now used to manage water levels for navigation, flood abatement, water supply, hydro generation and recreation. We saw many private boats cruising as well as many in the harbors along the way.

Our previous inland waterway tours have been on the English canals and on the "Mississippi Queen" and the "Delta Queen." The Canadian canals provide an opportunity to see a pretty part of Canada from a water point of view.

Information on the canal cruises can be obtained from:

Ontario Waterway Cruises
Box 6, Orillia
Ontario, Canada L3V 6H9
Phone (705) 327-5767

The "Kawartha Voyageur" enters a double deck lock near Healy Falls on the lower Trent-Severn Waterway.

OTTAWA LOCKS OF CANADA'S RIDEAU CANAL BEING REFURBISHED

By Bruce Russell - Contributing Editor

View of the 8 lock flight of the Rideau Canal at Ottawa, Ontario undergoing rehabilitation. May 30, 1993. (Russell photo)

The series of 8 locks at the Ottawa end of Canada's historic Rideau Canal is being extensively rehabilitated so that it will be able to function effectively for the remainder of this century and into the 21st. These lock chambers, arranged in a "flight" or sequence, provide a total lift of 80 feet so vessels can get from the Ottawa River into the canal situated at the top of the bluffs overlooking the body of water separating Canada's federal capital from Hull, Quebec. Constructed between 1826 and 1830, these locks fabricated from local limestone are the longest and highest in the Rideau Canal/Waterway system which opened for navigation in 1832. The Rideau Canal was the first artificial waterway in North America to be designed WITHOUT a towpath since steam power had been invented and use of mules to tow barges was considered redundant. Hence the Rideau was different than the well known American Canals such as the Erie, the C&O, and the Delaware & Hudson. The principle engineer in charge of building the Rideau Canal which connects Ottawa with

View of traffic in the Rideau Canal along the segment through Ottawa. This part of the canal was constructed through swamp and marshlands. Runners and walkers use the paths parallel to the canal to get exercise. (Russell photo)

Kingston on Lake Ontario, a distance of 124 miles, was Lt. Col. John By of the Royal Engineers of the British Army. Canada was then still considered "British North America" and wouldn't become a separate nation until 1867. The British at the time the canal was planned were still afraid that the newly independent American colonies to the south intended to invade Canada and forcibly join it with their country. Hence defense was a major consideration. The Rideau Canal, to be built along the route of an old Indian canoe passage from Ottawa to Lake Ontario, would permit waterbourne goods to be transported safely without the need to travel on the St. Lawrence River and through Lake Ontario where the US Navy might attack. As finally built the Rideau Canal is a waterway system which incorporates both man made channels as well as existing segments of the Rideau River which have been improved and converted into a navigable passage for vessels. It has 47 locks plus 22 blockhouses or mini forts for defense. However, Canada's fear of invasion by the Americans proved to be unfounded and in 1856 the British Army turned it over to the local Canadian authorities. It was used to move various commercial cargos including lumber until World War I when competition from railways made it superfluous. However it gradually became popular as a recreational waterway, a role it continues to fulfil.

The flight of 8 locks in Ottawa are not anchored in bedrock as are, for instance, the locks of the Erie Canal at Lockport, N.Y. At the time Col. By and his team of engineers were constructing them an attempt was made to dig into the sloping hillside deep enough to encounter solid stone upon which to place the sequence of 8 chambers. Unfortunately, the topography and geography of this particular location was such that bedrock was situated very deep, well below where the canal builders could reach using the tools at their disposal. Consequently they had to compromise and construct the lock flight on a less than solid rock foundation. In the 20th Century a problem such as this would be solved by driving piles into the ground deep enough to encounter solid material. Upon the piles the chamber would be erected. But in 1829 the powerful pile drivers of today didn't exist. Hence use of less efficient methods of stabilization were resorted to. Although Col. By probably knew what pile driving was, he was unable to drive piles deep enough to reach bedrock. Thus the 8 locks were constructed on existing surfaces.

On account of the lack of truly firm grounding, plus the effects of the harsh Canadian winters, by 1980 the flight of locks at Ottawa was in need of major repair and rehabilitation. Stones were cracked and each year fissures had to be filled. Original masonry had reached the end of its useful life. The Canadian government, aware that waterways such as the Rideau Canal and the Trent-Severn Canal were major national resources which brought tourist dollars to the Province of Ontario, authorized funding to do a major restoration on all eight locks. Beginning in 1983 local contractors, working under supervision of PARKS CANADA which operates and oversees all of the country's surviving 19th Century man made waterways, started the process of tearing apart the 150 year old locks. Worn and misaligned stones were discarded and replaced with new ones. Where lock floors had buckled or cracked new material was installed. In certain instances concrete was utilized to fill in spaces behind reconstructed lock walls, while leaving the visible surfaces cut and dressed stone. Hence nothing was changed outwardly. The restored chambers look as good as the originals but are from a structural viewpoint probably sounder and better equipped to deal with extremes in temperature. Col. By and his team of Royal Engineers would be gratified.

(Concluded on Page 70)

View of the Rideau Canal just beyond the top of the 8 lock flight in Ottawa, Ontario. The classic looking building on the right is the Hotel Chateau Laurier, a local landmark. (Russell photo)

RETAIN, RECORD, RECOVER, RESTORE, RE-CREATE
(And on the seventh day, nip over and see what they're doing in the US of A)

By (the late) Pat Saunders

I've spent the last five days (in January 1992) under the care of, variously, John Lindt, Vivienne Mitchell, Lynn and Mike Howlett, and (by telephone) Dave Johnson, viewing the C&O, Patowmack and Alexandria canals: Then with Bill Trout, George Rawls and Co., in Richmond.

This, to a British canal buff who joined the Inland Waterways Association as a young (-ish) ex-soldier in 1948, on returning from overseas service, was a great experience. We were all interested in the same sort of things, albeit against widely differing historical, social and geographical backgrounds.

In the UK in, say 1950, a whole canal system was still working in very much the same sort of way (with one exception), covering the same sort of routes, with the same sort of craft, using the same sort of techniques as in 1760, roughly the start of the canal revolution; a fossilized, but still working system. The one change was the introduction of the diesel engine, so that most, but not all, 30-ton narrow boats were propelled by an 18-20 hp diesel engine, instead of a horse. (Mules were rarely used, by the way.) A powered "motor" would tow an unpowered "butty," carrying between them 50-60 tons — it depended on the dredging. And this continued in the Midland Counties until the 1960's, when an un-typical cold winter, a three-month freeze-up, sent freight on to the roads, never to be recovered.

So, in some cases, it was campaigning for continued maintenance; in others fighting to retain at least the line of canal abandoned over the previous century, where recovery looked feasible; in still others, investigating and recording waterways gone (we had thought) beyond redemption.

My first physically active participation was in the early 1950's with work on the Lower River Avon; my last, to date, a few weeks ago on the Wiltshire and Berkshire Canal, near Wantage, fifty miles west of London, where over 1000 volunteers, organized by the Waterway Recovery Group, descended on one of those "no-hope" canals, and in the span of a weekend cleared 2 miles of towpath and canal bed of a ninety-year growth of weeds, shrubs, trees and rubbish.

It is the Waterway Recovery Group, a band of volunteers, of whom for many years Graham Palmer was the driving force until his untimely death, which provides much of the impetus behind the smaller, local, societies dotted over the country. Over the years, by its own efforts, linked with the IWA, it has become a source of experienced and skilled canal restorers with its own fleet of vans and minibuses, and equipment ranging from shovels through water pumps and chain saws, to dredging and earth-shifting machinery. Task forces travel hundreds of miles at the weekends and at holiday periods to "lend a hand," and work camps are organised right round the year.

Two recent successes, after year of slog, were the re-opening of the Kennet and Avon Canal, linking London with West Country Bristol; and the Basingstoke Canal, 30 miles of waterway SW of London. Her Majesty the Queen opened the former in August 1989 and the Duke of Kent the latter in May 1990. Both operations had taken two to three decades to achieve. I feel if some of us had known how long it would take — we'd never have started. Ignorance is bliss! There was one lady at the Basingstoke Canal who'd gone on her first work party lightly pregnant. Her *grandson* came to the re-opening.

My own home waterway, the River Wey, is only short, twenty miles linking Guildford and Godalming with the Thames SW of London. Then in the 1980's I saw and played on horsedrawn barges: it gave me the "bug." Horse drawn operation continued into the 'Fifties. Its last private owner, Harry Stevens, gave the Navigation to the National Trust, the principal British body caring for buildings and areas of national importance. On its 20 miles a thousand craft are registered, big and small: Recycling! Crowding!

The prime difference between our two approaches seems to me to be this: In the UK we are short of water to cruise on — so our aim is to "get it navigable and get boats moving on it" end to end. In the USA you have many miles of cruisable water so that incentive is less prominent and the accent is on the historical and archaeological aspects, with an environmental spin off. In Britain we have a Canal and Railway Historical Society which concentrates on that side. But many of us are equally happy steering a boat, dismantling an ageing diesel, delving into collections of documents or yellowing newspapers, raising money, hacking down encroaching vegetation or singing bawdy songs in canal-side pubs after a day on the "cut"; a sort of multi-faceted, three-dimensional approach. So in the US of A, so far, one element which seems vital to me is lacking — old canals working again. But who knows? In my sixties I can cruise on British canals I'd sadly regarded as lost for ever, at the age of twenty-five, even as an enthusiast.

As an Englishman I wouldn't dream of coming over here and telling you what to do. (Or would I?)

But I can dream, can't I? Dream of a manageable section of 19th Century waterway identified; with a support group raising funds, arousing public interest, pressurising the authorities, interesting boat-owners, learning old construction skills and using them; the aim to be the re-creation of a piece of American history which could be experienced by using it — and act as a revenue raising tourist attraction too. With, of course, the British Waterways Recovery Group being invited over for work camps.

Well, at 66 I doubt if I shall ever experience a 20 mile, 20 lock, 3 aqueduct day's cruise on a restored American canal. But my son might; or my grandson.

The essential quality of Brit canal restoration was this: we didn't know it was impossible, so we did it.

Keep up the good work!

(Pat Saunders was kind enough to write this article for American Canals *while on his way from England to Australia, stopping to visit family and canals on the way.)*

Rideau Canal

(Concluded from Page 69)

One of the excursion boats which travels the section of the Rideau Canal from the top of the 8 lock flight as far as Dow Lake, a 3 mile distance. (Bruce Russell photo)

Because the 8 lock "staircase" at Ottawa is not positioned directly over and anchored into bedrock it actually floats atop clay and gravel. Furthermore it's best to think of it as a series of inverted arches joined with one another. The restoration work took these factors into consideration and attempts were made to increase their stability by installing hidden stonework. Hopefully these improvements will prevent the locks from shifting position and developing cracks and fissures. The Ottawa flight is classic in appearance and is one of the most impressive examples of British and European lock design. Its setting in the heart of Canada's Parliament District adjacent to the Hotel Chateau Laurier adds to its appeal. Overlooking it is a bridge which allows passersby to observe vessels climbing the 80 feet from the Ottawa River to the upper level where an artificial channel begins. This passes through the heart of the city for approximately 3 miles before reaching Dow Lake. At the opposite end of this body of water another set of locks permits vessels to enter the next segment of the 124 mile long waterway. This portion of the Rideau Canal, from Lock #8 at the top of the flight, to the beginning of Dow Lake, was built at great cost in human life. The area was swampy and many of the laborers contracted malaria as they attempted to dig a channel through marsh. Eventually the Royal Engineers simply flooded the swamps and built retaining walls on either side of the water logged area.

During the warm summer months the section of the Rideau Canal in and around Ottawa is worth visiting. There are two companies offering short trips from the upper portion of the flight of locks as far as Dow Lake where they turn around. Boat trips are likewise available on the scenic Ottawa River. It's not possible to actually go from river to canal up the flight due to time considerations. If you desire to do this it's necessary to have or rent your own private boat. To negotiate all 8 chambers takes about 2 hours and even longer if there is a waiting line. However by walking alongside the flight on parallel paths it's possible to get the feel of passing through. A museum about the Rideau Canal is located in a historic stone building opposite Lock #4. Definitely worth a visit, it tells the complete story of the Rideau Canal from 1830 to the present, from the time of its military significance to its current role as a recreational waterway administered by PARKS CANADA.

"ACROSS THE SAHARA BY STEAMER"

Edited by Robert L. Trbovich

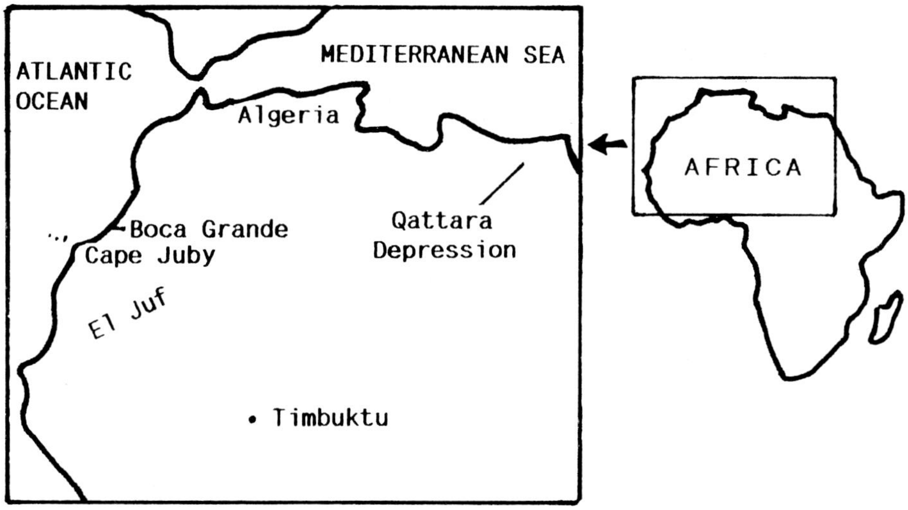

A century ago, Ferdinand de Lesseps and other visionaries planned to build canals into northern Africa to flood hundreds of square miles of below sea-level desert. The excitement of the time is clear in the following article by John T. Short in the July 1879 *Scribner's Monthly*. In more recent times, Israel had a plan to run the Mediterranean into the Dead Sea - see the February 1978 *National Geographic*, p. 230. What we would like to know is, what ever happened to these bold schemes? And was de Lesseps aware of the Qatarra Depression? Here are excerpts from the 1879 article, entitled "The Flooding of the Sahara":

"Across the Sahara by steamer" promises at no distant day to be classed among the announcements of this age of mechanical and engineering triumphs. That the Sahara, the figure in the world's literature and oratory for barrenness and solitude, should lose its old character and put on a new one in contrast with the old, would be surprising indeed. The proposal on the part of English and French engineers to flood the great desert with the water of the ocean and of the Mediterranean, and transform its waste into a watery highway for the commerce of the nations, at first impresses us as visionary and impracticable. A thorough examination of the subject, however, has led some of the most eminent scientists to quite a different conclusion. . . .

Probably the most original and feasible plan for laying open the heart of Africa is that recently proposed by both English and French engineers. It is no less than the remarkable suggestion which furnishes the subject of this article. Mr. Donald Mackenzie, the English engineer, proposes to flood the Sahara from the Atlantic, while Captain Roudaire and M. de Lesseps, the French engineers, are now engaged in work of a preliminary character with a view fo admitting the waters of the Mediterranean to the basins of Tunis and Algeria. . . . A large part of this salt waste is known as El Juf, the great sink of the desert, being 200 feet below the level of the Atlantic. . . .

Explorers agree that a channel once connected its north-western extremity with the Atlantic Ocean, at a point near Cape Juby opposite the Canary Islands. This channel, called by the natives, Sakiet El Hamra, presents abundant evidence of once having admitted the waters of the sea. Its bed is encrusted with marine salt, and shells. On the Atlantic coast it terminates in a sand-bank which prevents the waters of the ocean from flowing into its bed. Its great mouth, — Boca Grande, — resembling a miniature Gibraltar, is formed between perpendicular rocks which rise to a height of 200 feet above the sea; it measures between the rocks two and a half miles in width, and is blockaded by the sand-bar above mentioned, which itself has a width of 300 yards across, and a height above the sea of thirty feet at the south side of the channel, and ten feet at north. . . . Thus it is believed that El Juf was separated from the ocean and became successively a salt marsh, and a sandy waste, whose surface is still covered with abundant remains of marine life. . . .

Since the bed of the channel is at present 200 feet below the sea-level, the water of the sea could be utilized for the work of cutting the canal after a small ditch had once been dug across the bar. The torrent which would pour through the opening would no doubt soon remove the greater portion of the obstruction. However, when the great inland basin had again been filled and the water level both within and without the great mouth had reached an equilibrium, it may be presumed that the same agencies which gradually built up the bar, namely, the tides and waves, would renew their work, and, if left alone, would probably in time close the channel and again reduce the Saharan sea to a desert. Still, the influence of these agencies could in part be counteracted by the current which must flow out of the channel.

The volume of the rivers which now flow into El Juf would no doubt be greatly increased by the additional rain-fall incident to the presence of so large a body of water. The action of the tides and currents could be guarded by breakwaters and entirely controlled by locks. . . .

Were the Sahara flooded, and were it possible to steam from Liverpool or New York to Timbuctoo, the volume of trade would develop immensely. . . . Cape Judy is distant less than 800 miles from Timbuctoo. The depression of El Juf approaches within 100 miles of the city. The flooding of El Juf would consequently bring Timbuctoo within ten or twelve days' steam of Liverpool. Ultimately, no doubt, a ship-canal would be constructed between the sea of El Juf and the Niger.—a distance of 100 miles,—with which Africa will be thrown open to the world, and the "Dark Continent" will exist only as a dream of the past.

As previously intimated, the French engineer, Captain Roudaire, and the celebrated M. Ferdinand de Lesseps, whose energy and skill in the construction of the Suez Canal made Africa an island, are now taking steps with a view to flooding the *chotts*, or basins of Tunis and Algeria, from the Mediterranean. The exact extent of these depressions, if known, has not been publicly stated. The preliminary work of leveling, by Captain Roudaire, showed the depressions to be lower than the level of the sea, and demonstrated the necessity of making borings in order to ascertain whether rocks underlie the sand which composes the intervening ridges. At the meeting of the French Academy of Sciences, held December 9, 1878, M. de Lesseps gave an account of his visit to Tunis in company with Captain Roudaire and his expedition. . . .

The cheapness of labor in the desert will contribute greatly to the success of the enterprise. Bands of Arabs coming from all quarters have applied to Captain Roudaire for employment, asking but 90 centimes per day. Just how far the Tunisian sea will extend toward the center of the desert, cannot, at this stage of the enterprise, be conjectured. Whether it will be possible to effect a connection between it and the great depression of El Juf is unknown and doubtful. Should it extend as far inland as M. Roudaire contemplates making borings, namely, one hundred leagues, its southern limit will still be about one thousand five hundred miles from Timbuctoo, or twice the distance of Timbuctoo to Port St. Bartholomew.

The importance of flooding the Sahara and opening water communication with Soudan, from other than merely commercial considerations, cannot be estimated. In 1872, several authorities, among them Professor Berlioux of Lyons, estimated the annual export of slaves from Africa, including the destruction of life during transit, at upward of half a million. According to Sir Bartle Frere, the Superior of the Mission Convent of Central Africa fixes the annual (in 1872-3) drainage of human life from Africa, consequent upon slavery, at 1,000,000 lives. Lieutenant Young, writing from Lake Nyassa in 1876, describes the long lines of bleached human bones which he saw stretching toward the north-east, and states that the ground was covered with thousands of skeletons which mark he track of the slave driver. Is it not to be hoped that the opening of Africa would strike a blow, both final and fatal, to this horrible trade? Would not the introduction of a Christian civilization into this vast territory be one of the most magnificent missionary enterprises recorded in history?

ST. HELENA III COMMISSIONED

The christening and maiden voyage of the concrete *St. Helena III* lured a crowd of about 1000 spectators to Community Park, Canal Fulton, Ohio, May 2, 1992.

The Canal Fulton Heritage Society celebrated the boat's first day of operations by making three of the two and a half mile trips from the park to Lock 4 and back.

The long awaited event was the culmination of nearly three and a half years of work for the volunteers of the Heritage Society.

The boat was christened by **Peg Kilbride** with the help of her son **Randy Kilbride** who is vice president and was boat project manager.

Also on hand for the christening were **U.S. Rep. Ralph Regula** and **State Rep. David Johnson.**

(From the July 1992 Issue of The Canal Society of Ohio newsletter.)

NEW JERSEYANS TOUR IRISH CANALS

by "Captain" Bill McKelvey, Tour Leader

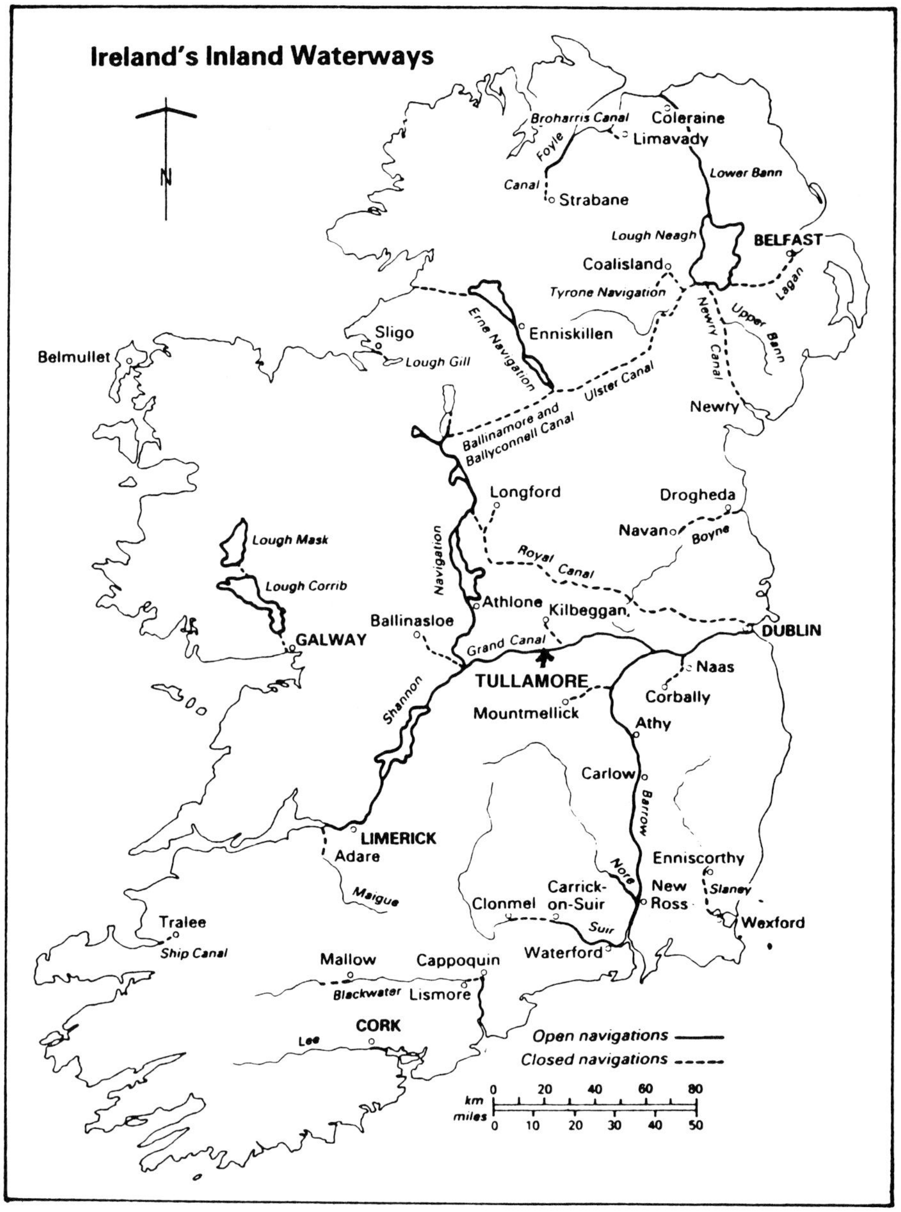

We are indebted to Ruth Delany for the above map, which forms the frontispiece for her IRELAND'S INLAND WATERWAYS, published in 1988; an excellent 200-page paperback, covering a 250-year celebration and history of all of Ireland's many canals, past and present. The book was obtained by J. William Shank, the Publisher's son, on a recent visit to Ireland.

A group of 22 members of the Canal Society of New Jersey arrived in Dublin on 17 May 1991 to begin a two-week study tour concentrating on the Grand Canal. Most of our crew members were experienced, having been on one or more of the five prior trips which the society had organized to the British Isles over the years. We must go over to the "old country" to experience the quaint old canals, as practically none of the original towpath canals in the United States are available for pleasure boating.

From the airport, our coach took us on a sightseeing minitour of Dublin, then on to our first stop—the Guinness Museum and Pub. It provided a fine introduction to the brewing process. We were most intrigued by their Transport Gallery (which included a video on the Guinness links with the Grand Canal) and, last but not least, in sampling a pint or two of their outstanding brew. All had the balance of the day on their own. Some of us took a ride on the Dublin Area Rapid Transit (DART) system and discovered a pleasant fishing village on the north end called Howth. That evening, we were given a lovely reception by the Dublin Branch of the Inland Waterways Association of Ireland (IWAI), along the canal at Portobello. We had gotten our introduction to the famed Irish hospitality—more was to follow.

The next morning, Saturday, after seeing that all our people were safely aboard the train for Tullamore, I strolled up to the cab and asked the engineer if I might ride with him. His response: "Why sure, my lad, climb aboard." It was a ride not to be forgotten, for the engineer recited appropriate passages from the works of several Irish poets as we cruised along at 90 miles per hour. One of his favorite passages was, "Be sure, the great God never planned for slumbering slaves a home so grand." He loved his job and his country.

Celtic Canal Cruisers transferred us from the station to their base at Tullamore. Our five boats were ready and we were given a thorough orientation. We were delighted with the roominess and arrangement of the 9-foot-wide boats which we had chartered. Shortly we were on our way east, tying up at Daingean overnight. It was a small town of about 30 buildings, but with 11 pubs.

On Sunday, the 19th, we cruised to Robertstown, through countryside much more rural than we had been used to in the United Kingdom. At this stop, the Kildare Branch of the IWAI presented each of us with a beautiful brass plaque and a map of the Grand Canal system at its peak. It served the ports of Waterford, Limerick, and Dublin, as well as 19 counties. Later that evening, we met with a number of their members at a canalside pub where they arranged to have Patrick J. Kane, a retired canal boatman, on hand for us to chat with. Stores for food supplies and pubs (restaurants) were conveniently located at nearly all of our overnight stops.

We headed down the Barrow Line canal on Monday morning, passing several old steel boats, once used for freight, tied up at Lowtown, and met the dredge en route to a new job. At first we thought it was sinking, but then discovered it was designed to travel with its forward deck awash. This apparently made it easier to get under bridges. We were impressed with the level of maintenance of the canal prisms—good width and depth throughout. We decided to take a hike up the Milltown Feeder, but discovered to our dismay that we had chosen the wrong bank, as the path became covered with dredged material.

We enjoyed our stops at Rathangan, Monasterevan, and Vicarstown, the latter being our turnaround point. Monasterevan was particularly interesting due to the mainline railroad crossing plus old warehouses, an aqueduct, and a unique lift bridge. There were many photo opportunities! Jim Ridgway of the IWAI Offaly Branch advised us to "travel slowly and stop frequently." We did, and we loved it. Jim was particularly helpful with suggestions, information, and guidance.

Next, we backtracked to Lowtown and headed west on the Grand Canal, stopping for an afternoon in Edenderry Harbour. I decided it would be better sleeping outside the center of town, so my boat cruised east and tied up at the Board na Mona (Irish Peat Board) narrow gauge railway bridge. This allowed us an opportunity to explore the harvesting of turf (also called peat) which is burned as fuel in power plants, and to see the machines, equipment, and railway operations.

Dredge on the Barrow Line canal. (Photo by Robert H. Barth.)

New Jerseyans negotiating unidentified lock. (Photo by Robert H. Barth.)

On our return to Tullamore, we tied up in the old harbour where the canal engineering depot is located. An inner basin graveyard contained numerous old steel commercial and maintenance vessels which were mostly sunk, perhaps to retard further deterioration. We were permitted to inspect the shop where new lock gates were being fabricated. The material used was Ekke wood from Burma—very expensive but extremely dense and long lasting in canal service. The quality of repair and maintenance work which we observed in process there and at several locations was, in our opinion, very high. Proof of this was that all locks and bridges were found to be in excellent working order. In contrast to United Kingdom canals, nearly every lock we used was readied for us by a locktender, although we usually helped.

Through the fine hospitality of the IWAI Offaly Branch, a day-long, most interesting itinerary was set up for us on Saturday, the 25th. We boarded a bus and first toured Clara Bog, preserved in its natural state, then on to Locke's Distillery Museum, where we had a three-course lunch. In the afternoon, we toured Kilbeggan Harbour at the end of the disused Kilbeggan Branch. It was good to see the ambitious restoration in process and to learn of their plans to reopen the branch to navigation. It made us envious. Lastly, we visited the Irish Mist Offaly Heritage Center, and were given samples and souvenirs. We were honored to have Paddy Egan as our guide for the most pleasant day.

The next day we resumed our westward trek, passing castles, mills, pretty farms, and more active peat works, which were part of the immense Bog of Allen. One farm had a fording location where the cattle crossed with the canal water up to their necks. At another location, boys were jumping off a railroad bridge into the canal. Another thing which we found quite different from our experiences in the United Kingdom was the near absence of other boaters on the canal. We usually had the canal to ourselves, with no traffic jams at locks. The largest number of boats we saw at one place was at Shannon Harbour. It was a most interesting collection, with many former cargo boats which had been converted to pleasure use. It was sad to see the old canal hotel there in such a deteriorated state. We hope it will be restored.

For the next several days, we adopted a different style of cruising for the placid Shannon River. We saw all kinds of pleasure boats and frequently passed them. Many were hired by vacationers from other countries, usually German. First we headed north and tied up in a channel between an island and the riverbank, just below the town of Shannon Bridge. This town is blessed with the most hospitable, friendly people, and more of them than at any of our other stops. One shopkeeper suggested that the power plant might give us a tour. The next morning, we marched up to the gate in a group to ask if they would let us in. Two employees were dispatched to guide us throughout the entire peat-burning plant. They gave us coffee and cookies, and passed out literature and art prints as we departed. We would like especially to express our thanks to Derry Killeen and Patrick Dooley.

We Take a Train Ride

Later that day, we took a special tour train ride (on the Clonmacnois and West Offaly Railway) through the turf-gathering operation for the Shannon Bridge power station. We all came away from this experience with a thorough knowledge of the process, from the creation of a bog to the disposal of the ash residue. In the afternoon, we continued our cruise up the Shannon to Clonmacnois, the remains of a 1,000-year-old monastery in a most scenic setting. We tied up there overnight.

The next day we continued north to Athlone, a fair-sized city, which we thoroughly explored, and then headed back down the Shannon, tying up again at Clonmacnois. On the following day, we continued south to Banagher, where we tied up at a convenient boat base. So far, the weather had been absolutely delightful. On several days, there was not a cloud in the sky! That evening, we went to a famous, quaint, old pub called "Dirty Nelly's," where we were entertained with traditional Irish music by local performers.

Our most southerly excursion on the Shannon was to Meelick Lock on the next day. None of us had ever seen wood mitre gates this large before, and these were bowed as well as hand operated. On our way back, opposite the entrance channel to the Grand Canal on the east side of the Shannon, we could see the first lock of the Ballinasloe Line, the derelict westward extension of the Grand. Not to worry, this canal is also slated for restoration. And that is not all—it is estimated that the Royal Canal restoration will be completed in five years; the Royal will be linked to the Grand; and the north and south of Ireland will be linked as well. If only we could get such activity going in the United States.

Debris in the Canal

Alas, our tour was coming to an end, and we all had to head back to Tullamore. The water of the Shannon had been clean and clear of debris—not so the canal. Tough plastic bags, sheeting, and baling twine (mostly from farms) seem to be attracted to the canal water. Some of this seems to float half way between the surface and the bottom, just waiting to snag our spinning propellers. We were back to our periodic routine of reaching down into the weed hatch to remove the entangled mass, sometimes more than twice daily. This material simply does not biodegrade! While I am on the subject of needed improvements, there are two more which can be added; better facilities for getting on and off boats below most locks, and posting of identification numbers at each lock and bridge.

On a positive note, we were very pleased with the guides to both the Grand and Royal canals, as well as the Inland Waterways News, all published by the IWAI. Information on their availability can be obtained from Ruth (Delaney) Heard, Stone Cottage, Claremont Road, Killiney, Co. Dublin, Ireland. In addition, a series of Canal Bank Walks guides is available from the County Kildare Sports Advisory Committee, and several books have been published in the past.

By late Friday afternoon, May 31st, all of our boats were back at the Tullamore base. The last item on our agenda was our traditional Canal Society of New Jersey smorgasbord. This one, beside Lock #24, was one of our best ever. Joining us were Paddy Egan, Mike and Heather

(Concluded on Page 74)

The challenge of upward mobility. (Photo by Dennis Forbush.)

IRISH CANAL TOUR

A narrow-gauge peat-hauling train "navigates" through Kylemore Lock and bridge in the prism of the abandoned Ballinasloe-Shannon Harbor Link of the Grand Canal. Note the remains of the wood mitre gates.

(Concluded from Page 73)

Thomas (proprietors of Celtic Canal Cruisers), and the locktender. Just as we were finishing, a group of several canoes came through the canal on a fund-raising marathon.

On Saturday morning, we were driven back down to the railroad station for our return trip to Dublin. That afternoon, by prior arrangement, Ruth Heard picked up three of us at the hotel and gave us a delightful tour of the many points of interest along both the Grand and Royal canals in and around Dublin. Who better to lead us than the author of several books on Irish waterways and the editor of the IWAI News? It was most enjoyable.

We were most impressed with the excellent condition of the canals and the work that had been completed toward restoration of navigation. We saw where bridges had been provided over the canal for a new motorway to maintain the canal right of way. This was certainly a change from what we see in the United States, where canals have been covered over for new highway routes.

While waiting for our plane at the airport, we saw the first precipitation since we got on our boats 15 days before. I guess the farmers were happy to see us go, for they really needed the rain. As if having a great time was not enough, participants were delighted with the sizable rebate checks they received after all accounts were settled. The entire trip, including air fare, boat rental, bus trips, two nights in a hotel, and some extras, was executed for just a shade over $1,200 each.

We look forward to the future when the Royal Canal is reopened for through navigation. We will return to do it!

SUPPLEMENTARY NOTE: Peg and Zip Zimmerman stayed on to tour Ireland by rental car for nine days after the conclusion of the tour. Among other things, they discovered a unique bed-and-breakfast establishment at Mullingar on the Royal Canal. Lock 29 is soon to be restored, and the lockhouse is being converted to a bed and breakfast. There are many lock sites where you can tour the lockhouse, but this is surely one of the few where you can sleep in it. For information and reservations: Mrs. Lorraine Hornidge, Lockhouse #29, Mullingar, Co. Westmeath, Ireland (phone 044-73157).

Here the rail line passes through the arch of an old canal bridge. The Irish Peat Board operates 800 miles of such railway to transport the fuel to power plants. Similar lines cross over the Grand Canal main route at several points on moveable bridges.

WEDDING OF THE WATERS REENACTED

On the 26th of October, 1825, the Seneca Chief left Buffalo for New York City on a trip that was to be the official opening of the Erie Canal. On board were two kegs of Lake Erie water, to be poured out into the Atlantic Ocean at New York Harbor in a symbolic "wedding of the waters" being joined together by the new canal.

A few of the details were different, but the essence of the ceremony was reproduced in 1991. The tug Lockport started from Tonawanda on the 12th of October with its shipment of Great Lakes water. At the same time, New York voters were preparing to go to the polls to vote on a proposal to reenact another feature of the original Erie Canal—canal tolls. (The proposal was passed on November 5th, 1991.)

BOOK REVIEW

Harry L. Rinker, *The Schuylkill Navigation: A Photographic History* (Berkeley Heights, N.J.: Canal Captain's Press, 1991).

Reviewed by Albright G. Zimmerman

This reviewer joins the myriad of canal buffs who always view any new offering dealing with canal history and lore with great anticipation. The recent publication of Harry L. Rinker and Bill McKelvey's Canal Captain's Press easily satisfies that anticipation. It is a well-conceived and well-implemented, much-needed pictorial history of the Schuylkill Navigation. There is a wealth of well-selected pictures, and on the whole the texts under the illustrations are adequate. The historical sketch, while brief, is comprehensive and essentially accurate. The inclusion of "Folklore of the Schuylkill Canal," by John Bowman, is a stroke of genius. The bibliography is both extensive and exhaustive. This book should find a place in the library of every serious canal scholar and of every canal buff.

Having thus bestowed lavish and well-deserved praise, I can now take this opportunity to make some general comments about what might be hoped for in canal literature, and at the same time to satisfy one of the seeming obligations of a reviewer, nitpicking at the work being evaluated. First, let's consider what this book is and what it isn't. While it more than adequately fills the need for a pictorial recording of the past of one of our canals, it does not answer another real need, that is, for detailed, researched histories, both popular and scholarly, of our canals, something that begs ever-new commitments, and something that has been better achieved on other canals and in states other than Pennsylvania and New Jersey. It should also be remembered that there are a number of possible thrusts for such works, such as emphases in areas like political, economic, technological, and social history.

The producers of works such as this should subject their data—in this case, their pictures—to a much more rigorous critical evaluation. Rinker, in his acknowledgments, bruised some egos in his critique of the documentation of photography in another well-conceived and important work, yet I must make some suggestions for enhancing the value of the work under review in this same respect. I hope that these will be taken as constructive criticism by Rinker and, further, as a set of goals to be kept in mind by others embarking on similar works. From my own experience, I know what I ask is difficult to implement, in many cases impossible, but every author-editor-compiler and, yes, collector, should attempt to date and otherwise document every picture.

The volume might better have led off with the earliest representations of the canal, the concluding Staffordshire illustrations, presumably adapted from Thomas Birch's 1821 painting but with romanticized additions. The original Birch painting might well also have been included, or, better yet, the possibly earliest pictorial representation of the Schuylkill Navigation, an 1824 sketch of the dam at Fairmount showing the first locks on the Schuylkill Canal, found in the 1832 American edition of the *Edinburgh Encyclopedia*.

Despite the foregoing criticism, which come partly from my prejudices as a professional historian, I shall treasure my copy of the book. I recommend it to all members of the wide fraternity of canal buffs.

The Schuylkill Navigation can be purchased from Canal Captain's Press, 103 Dogwood Lane, Berkeley Heights, N.J. 07922, for $16 plus $2 postage and handling charge.

MAIN-DANUBE CANAL OPENS

The big news in Europe is the official opening of the 106-mile Main-Danube Canal through the beautiful Bavarian countryside in southern Germany, September 26, 1992. This canal, completed after nearly thirty years of work on the part of the German government, is the final link in the system of major inland waterways across Europe to connect the North Sea with the Black Sea. Large barges carrying 2400 tons of bulk cargo, not to mention sizable passenger ships, will now be able to travel from Rotterdam to Istanbul, without taking the long voyage around France, Spain, Italy and Greece in the Atlantic and Mediterranean. The entire inland system, involving the Rhine, Main and Danube Rivers, with full canalization where necessary, has been planned and under construction intermittently since 1921. For many it is a "dream come true". (For others it is a threat to the environment of central Europe.) For the first time, river ports on the Rhine River will be directly accessible to river ports on the Danube, by water.

The Main-Danube Canal climbs from Bamberg, close to the upper reaches of the Main River, over a summit 1332 feet above sea-level, higher than any other major waterway in Europe. It utilizes eleven huge locks up the north-west side of the summit and five locks down to Kelheim on the Danube, on the south-east side. The locks are 39 feet wide, 623 feet long and some have a lift of 100 feet.

For a full description of this modern miracle of canal engineering read the August 1992 Issue of National Geographic, pages 3 through 31.

Chairman's Report

The Deutscher Kanal- und Schiffahrtsverein Rhein-Main-Donau e.V. (German Rhine-Main-Danube Canal and Shipping Association) has been committed for more than 90 years to linking the divergent rivers Rhine and Danube. This goal is now near at hand; the Main-Danube canal is scheduled to reach Kelheim in1992, which will close the last gap in the route between the North Sea and the Black Sea. This waterway will then link up 13 European countries along the courses of the Rhine, Main and Danube. The German Canal and Shipping Association was founded on November 6, 1892, in Nuremberg. Its members include 29 cities and municipalities, 13 chambers of commerce and other commercial institutions, as well as 286 firms and individuals — an extensive forum for publicizing the age-old concept of a link between Main and Danube.

As protagonist of the complicated and protracted project of the Main-Danube waterway, the Association has had to face up to critics again and again and to ward off attacks against the project with sound arguments. Even now, shortly before completion of the final stretch between Nuremberg and Kelheim, this situation still prevails. While in the initial phase the engineering planning pushed ahead by the Association played the major role, its current activities are distinguished by two studies initiated and substantially financed by the Association. At the end of the nineteen-sixties, when some uncertainty became apparent at home and abroad as to the future legal status of the Main-Danube Canal, the Association commissioned a study of the status of the future Rhine-Main-Danube waterway under international law. The findings of that study, i.e. that the Main-Danube Canal is a national waterway, have meanwhile met with general acceptance.

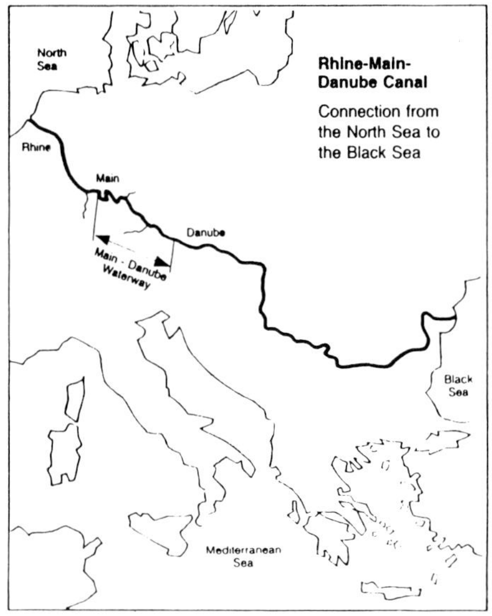

Construction on one of the final, level sections of the Main-Danube Canal, below the Berching Lock. (Photo by Roger Squires, May of 1990.)

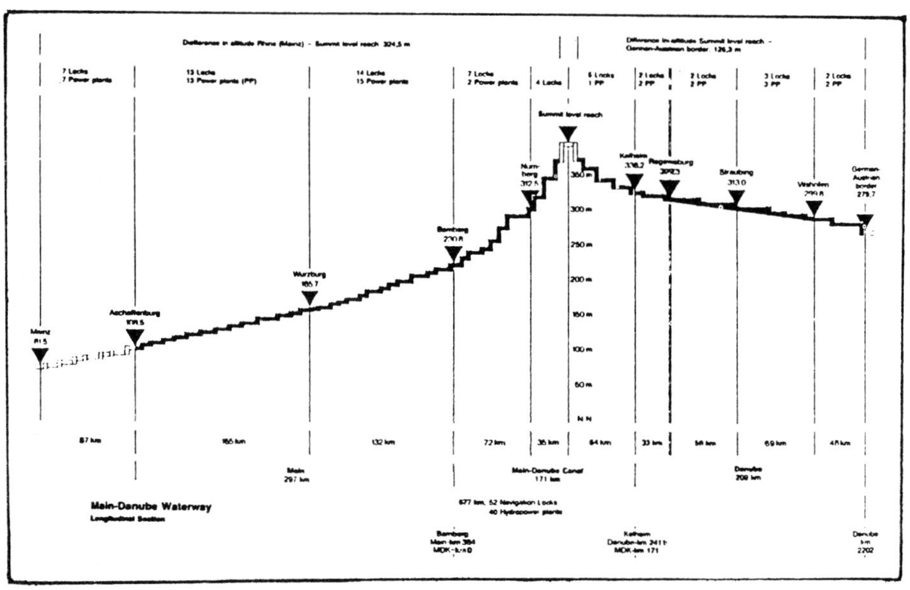

An elevation chart of the many locks on the Rhine-Main-Danube navigation system, built over the past seventy years.

In the early nineteen-eighties, when the controversy between advocates and opponents of the canal became ever more confusing, the Association retained the Ifo Institute, Munich, to analyse, evaluate and assess the factual content of arguments and counterarguments.

This study of the Main-Danube Canal clearly established that the waterway has a positive cost-benefit ratio and should be completed if only on account of its advanced state of construction.

Since the once-in-a-lifetime project of a waterway connecting Main and Danube is attracting increasing international attention, it seems fitting to publish a description in English. It is presented here by the German Canal and Shipping Association, with the kind assistance of the Rhein-Main-Donau AG and other institutions.

Chairman of the Board
of the
Deutscher Kanal- und
Schiffahrtsverein Rhein-Main-Donau e.V.

THE REMARKABLE CORINTH CANAL

By Bruce J. Russell - Contributing Editor

Each year over two million Americans travel to Europe and among them are canal and inland waterway enthusiasts. While the English and to a lesser extent the French and German canals are the prime attractions, Europe has other waterways which are also worth visiting. One of these is Greece's CORINTH CANAL now approaching its centennial.

The ISTHMUS of CORINTH is a 4.1 mile wide sliver of land connecting the mainland of Greece with the Peloponesian Peninsula. On its western side is the Ionian Sea and on its eastern flank is the Aegean. Since antiquity it has served as a natural obstacle preventing the passage of ships from one to the other. Here vessels powered by sails and oarsmen (often slaves) were forced to dock at either side of the isthmus and unload their cargoes which were subsequently hauled overland by animals, slaves, or a combination of the two until the opposite shore was reached. The goods were then reloaded onto other ships and the journey resumed. By 500 BC the Greek city of KORINTHOS or Corinth had become wealthy by providing men and materials for this essential service. Taxes and other levies were imposed upon ships docking at its port, and during countless wars it was a military objective. Control of Corinth meant control over a substantial share of the ancient world's commerce.

In 135 BC the Romans conquered Greece and turned their engineering skills to the situation at Corinth. The Romans initially constructed a road across the isthmus over which complete ships mounted on giant rollers could be moved. An abundance of slave labor provided sufficient manpower to move the wooden ships from one shore to the other. It must be remembered that the ships of antiquity rarely exceeded 200 feet and were made of wood held together by iron clamps. Thus the process of moving one might easily be accomplished during the course of a day even if the rollers traveled at the rate of only a half or quarter mile per hour.

The Roman emperor NERO in AD 67 made the first serious proposal to excavate a ship canal at Corinth to connect the two seas. History records that a team of engineers studied the situation and actual work commenced. Unfortunately for the Romans the Corinthian isthmus consists of solid rock beneath a light cover of topsoil and the invention of dynamite and blasting powder was over 1500 years in the future. Nevertheless they did succeed in excavating some stone and their cuttings remained visible until modern times. They were ultimately obliterated once work began on the present waterway.

During the 1600s and 1700s as ships became larger the practice of transshipment became impractical and the alternative of sailing an extra 210 miles around the Peloponesian Peninsula became accepted practice.

Modern Greece evolved in 1825 following independence from the Turkish Ottoman Empire, after a bloody struggle. Once the new nation became organized thoughts turned to internal improvements. Foreign engineers - French, German, and British - were invited to come to Greece and study the problem of breaking through the 4 mile wide ISTHMUS of CORINTH. Most shook their heads and said it could not be done. In their opinion the sheer volume of heavily faulted limestone which had to be removed was prohibitive. By 1870 nitro glycerine, black powder, and dynamite had been invented and all were being employed to construct long railway tunnels beneath such formidable mountain ranges as the Alps in Switzerland. The impossible slowly began to enter the realm of feasibility and the Greeks pressed on with their notion of a trans-isthmus waterway.

Because of the swift, west-to-east flow of water in the Corinth Canal all ship transits must be done with pilots and tow-boats, as shown here. (Photo by Spyros Spyrou-Corinth.)

In the mid-1870s a decision was finally reached to proceed with digging a sea level canal between the GULF of CORINTH and the SARONIC GULF. This was the narrowest point and was, according to historians and archeologists, where the Romans used to move their vessels using rollers. Foreign engineers as well as Greek ones determined that one of the biggest hurdles would be physically carting away the debris of stone after blasting had taken place. Millions of cubic feet of rock and rubble would have to be removed once the dynamite and nitro glycerine had done their work.

Fortunately for these late 1870s canal builders railroads and steam powered shovels had been invented. By spiking down temporary tracks it was possible to shovel the residue from each round of blasting into railroad cars and have trains take it away to be used elsewhere as fill. The French designed construction locomotives and side dump

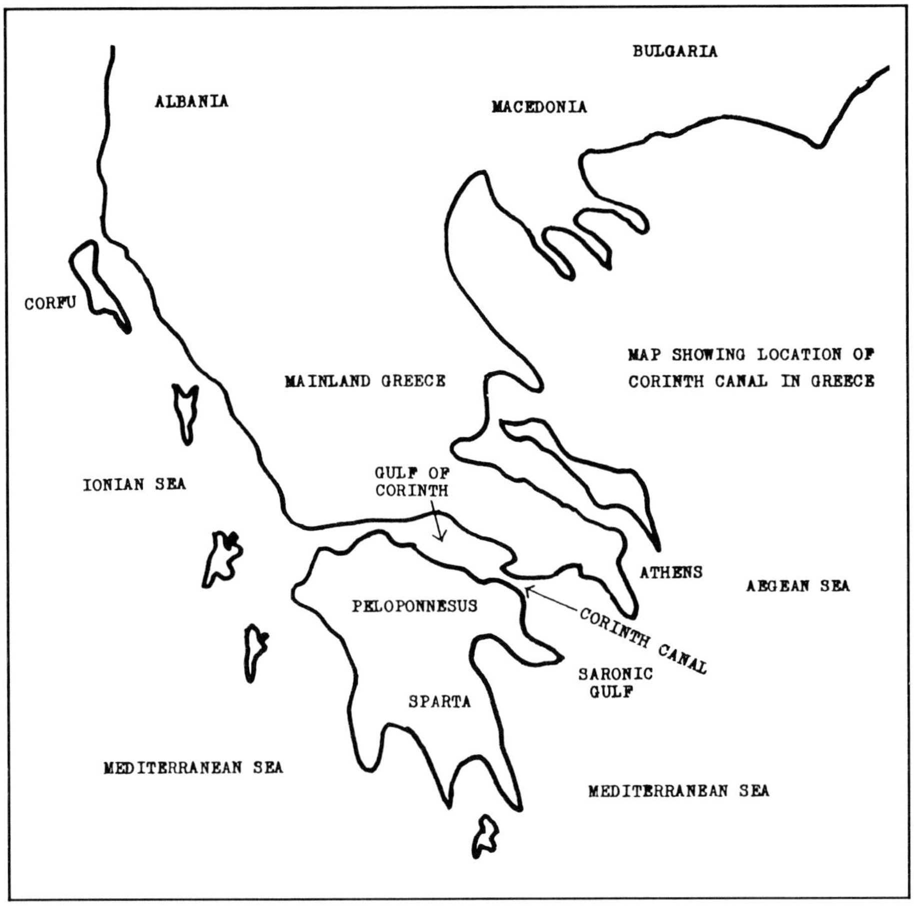

cars were sent to Greece and the work of cutting a canal through 4 miles of stone commenced.

For 13 years from 1880 until 1893 the herculean project moved steadily forward. Tourists visited Corinth to view the activity from special observation platforms. The fragments of pulverised limestone removed from what was to become the waterway were used to build countless harbors, jetties, and other civil engineering improvements throughout Greece. Once the man-made ditch had reached sea level small walls of rock at either end were left in place to serve as coffer dams while excavation penetrated another 26 feet below sea level. This was to create sufficient clearance for the largest ocean going vessels of the day. By 1880 steam had replaced most, but not all, sailing ships. Since the walls of the Corinth Canal were almost 300 feet high there was never any need to take into consideration clearance problems for the masts. The first bridge to span the canal was for a narrow gauge railway and later for a highway crossing.

When the CORINTH CANAL finally opened in 1893 it was hailed as a wonder of the world. People came from all parts of Europe to marvel at it. The final dimensions were 81 feet wide, 26 feet deep, and 300 feet high at the point of greatest elevation. During the years of construction the French company which held the original concession from the Greek government encountered financial difficulties and went bankrupt. A Greek consortium completed the job. Many visitors succombed to dizziness when looking down its straight rock walls, but all departed greatly impressed at the remarkable feat of engineering. Since the major portion of Greece was now geographically severed from the Peloponesian Peninsula the latter technically became a huge island. Furthermore because the Gulf of Corinth is about 51 centimeters higher than the Saronic Gulf there exists a permanent and swiftly moving west to east flow of water. This creates hazardous conditions for small boats.

Once it opened, the CORINTH CANAL provided a significantly quicker route for Italo-Greek shipping. Furthermore it became an inspiration for engineers in other nations who proceeded to excavate canals at Kiel in Germany and at Manchester in Great Britain. In the early years of the present century minor improvements were made such as deepening it and installing night time illumination. The greatest threat was always considered to be an earthquake but fortunately none have struck in the vicinity of the waterway.

Construction activities on the Corinth Canal, 1882-1893. Rail cars were used to service the digging site and carry the excavated rock away.

The CORINTH CANAL along with the rest of Greece fell into German hands during World War 2. The remains of concrete pill boxes housing anti-aircraft batteries are still visible at various points alongside of it. Both the Germans and Italians used it to send ships laden with men and material to the various fronts and it was a natural bombing objective for Allied warplanes. Their goal was to disable and if possible sink a ship during its transit, thereby rendering the channel useless. Allied commandos likewise considered planting dynamite charges inside its walls and detonating them by radio. Fortunately World War 2 ended before such a scheme could be implemented.

Following ending of hostilities and a return of prosperity during the postwar years pleasure cruising began to develop in Greece and the eastern Mediterranean. The sailing plan of many Aegean and Ionian cruises now includes a passage during daylight hours through the CORINTH CANAL, regarded by many as the most spectacular canal ever built with its towering and almost vertical rock walls. Cruise ships entering the waterway from either side are pulled through by tugboats whose captains are familiar with the often unpredictable west-east flow of water. Meanwhile, 300 feet above, tourists stand on the sides of a highway bridge and marvel at the ships making their transit.

American canal enthusiasts, if planning a tour of southern Europe, should inquire about trips through the CORINTH CANAL. Several moderately priced 3 and 4 day cruises include it in their itinerary, especially if they are heading from Pireus to the island of Corfu adjacent to Albania. The Greek National Tourist Office in New York City can supply information. If time for a nautical journey is not available take a short train or bus ride from Athens to Corinth and spend time watching ships pass. During the busy summer season continuous activity is assured, and during the remainder of the day Corinth's ancient ruins and archeological sites can be visited.

CANADIAN HAPPENINGS

John Jackson and Sheila Wilson's latest history of St. Catharines, *St. Catharine's — Canada's Canals City*, was recently launched at the St. Catharine's Museum at Lock 3. While not solely a canal history, it contains a liberal amount of canal data and illustrations. Another book, intended as a companion piece to their extremely popular *The Welland Canals — The Growth of Mr. Merritt's Ditch*, is Roberta Styran and Rob Taylor's *Mr. Merritt's Ditch — A Welland Canals Album*. Both books are 'musts' on the bookshelves of canal enthusiasts. The authors have selected a wide range of photo, illustration, and map material to illustrate their latest book, as well as having undertaken new research which is literally rewriting some of the past thoughts on the Welland Canal. The book was officially launched on Canada Day, July 1st, at the Welland Canal Viewing Complex at Lock 3 in St. Catharines. And lastly, as part of a special series of meetings with business and industry, Seaway President Glendon R. Stewart gave his views on the future of the St. Lawrence Seaway. The event suitably took place on August 6th, in St. Catharines, the 60th anniversary of the Official Opening of the Welland Ship Canal.

Arden Phair

JAPAN'S BIWA CANAL: ALL-PURPOSE WATER UTILIZATION

By Gordon J. Thompson (Edited by David Ross)

Several factors deterred navigation-canal development in Japan: the nation's late entry into the industrial era (Japan went right to work with railway development, having missed the canal-building era); topography typified by steep-sided mountains and short, rushing streams; and few shallowly-sloped inland river valleys. In addition, this island nation surrounded by navigable seas lacked incentive to develop inland canals. Water transportation was and still is more important in Japan than it is in most countries, but it is nearly all coastwise shipping. Nevertheless, a variety of canals did develop. In the coastal plains that accommodate the vast conurbations of Tokyo-Yokohama (the Kanto Plain) and Osaka-Kobe-Kyoto (the Kansai Plain), the Japanese widened and deepened streams and dug additional connecting links to create networks of sea-level waterways to bring goods from the shores into the cities. In a few inland locations, they canalized rivers by straightening, diking, and dredging. In the fortified cities, waterways were excavated to bring water to fill moats surrounding castles. For none of these were unique types of boats devised. To this day, man controls water through an extensive, comprehensive public works system of canals that channel water to rice paddies and to industries. These, still built anew or improved from time to time, often appear like North American canals, but navigation does not occur, and the bridges usually clear the water by less than a foot.

In only a few locations did the Japanese build the type of canal familiar to North Americans and Europeans to create cross-country waterways in directions not intended by nature. Remarkable among these is the 13.32-mile Lake Biwa Canal/Keage Incline facility near Kyoto. It is reminiscent of the C. & O. Canal, with its Paw Paw Tunnel, and the Morris Canal, with its inclined planes. No longer used for navigation, it continues in use and good repair for the other functions for which it was designed.

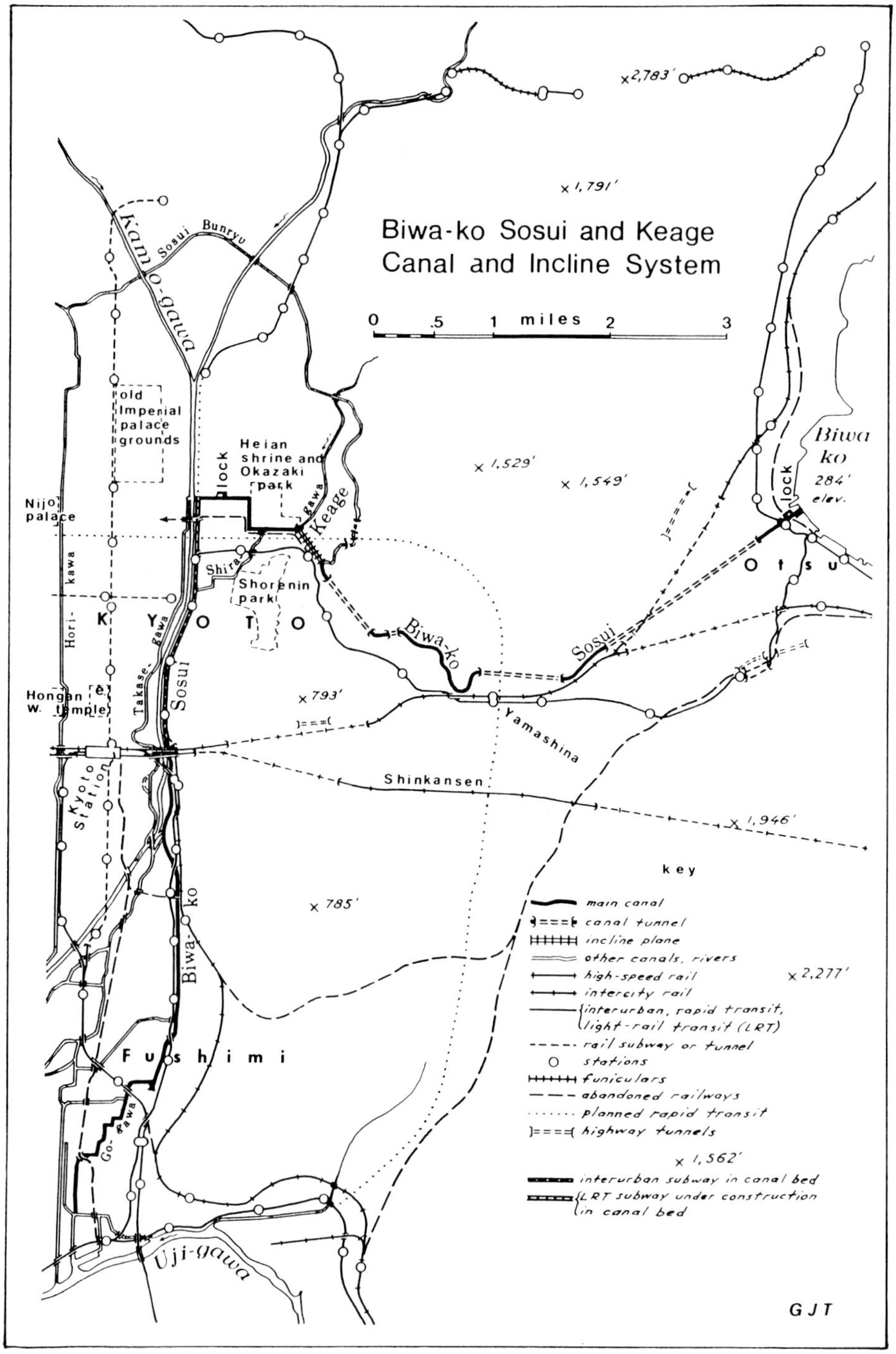

Looking west from the pedestrian bridge to the east portal of the third tunnel.

The Setting

According to tradition, nature created Biwa-ko (ko = lake), Japan's largest lake, with a mighty thrust of the earth's thin mantle when Mount Fuji erupted in June of 286 B.C. The size of Lake Geneva in Switzerland, 265-square-mile Lake Biwa drains a basin surrounded by mountains. With a surface elevation of 284 feet above sea level, it ranges as deep as 318 feet. Uji-gawa (gawa = river) drains the lake, dropping south, southwest, and northwest over an indirect course passing south of the city of Kyoto on its way to the Inland Sea at Osaka. A tributary river, the Kamo-gawa, enters Biwa River beyond Kyoto after passing through the heart of the city.

The lake's basin slices the volcanic spine of Japan's home islands. Forming a natural barrier across the large island of Honshu, for centuries it served as the frontier between civilization and wilderness occupied only by the aboriginal Ainu. This vast water supply lies tantalizingly close to Kyoto, but separated from it by the Hiei Sanchi Highlands — high ridges containing many peaks. Nature carved a narrow path through which the Uji River flows. This defile, indirect and containing canyons and waterfalls, is not conducive to transportation.

Kyoto, an old city, served as capital of feudal Japan from A.D. 794 to 1868. A cultural and religious center with no heavy industry, Kyoto was spared from bombing during World War II because of its palaces, shrines, temples, art, and other antiquities. The light industry that exists is devoted to textiles, ceramics, and lacquerware, and is largely carried out in homes or very small establishments. After the Japanese established Tokyo as the new capital, commerce between the two cities began to grow, with the well-trampled path taking on great importance as a national road—the Tokai-do (Tokai = east seacoast; do = road) — that now lends its name to the populous coastal conurbation between Tokyo's Kanto region and Osaka's Kansai region. Samurai warriors, other travelers, and later ox carts journeying the Tokai-do found their way over the Hiei range through a steep course and the Otani Pass—a low point of about 574 feet elevation, second only to the Uji River's gorge. Otani was undoubtedly an ancient outlet for the lake. Over time, the city of Otsu grew as a stopping place on this national highway at the outlet of Lake Biwa into the Uji River.

After 15 years of construction, in 1879 an early railroad pierced the barrier over an indirect route making partial use of the valley of the tributary Yamashina River, but resorting to steep gradients, heavy cutting, and the pioneer Osakayama Tunnel, now a historic landmark. This formed a vital link in the Tokaido Hon-sen mainland railroad from Tokyo through Yokohama, Nagoya, Kyoto, and Osaka, to Kobe.

The entrance to the Lake Biwa Canal, looking east from a street bridge in Otsu. Lake Biwa is in the background. The guard lock shown is apparently no longer in use; the bridge abutment is from an abandoned railroad.

The west portal of the third tunnel. Gates to the reservoir and water treatment plant are shown at right.

The Biwa/Keage Project

In 1881 (14th year of the era of progressive Emperor Meiji), Kitagaki Kokudo, then Governor of Kyoto prefecture, asked a university engineering student to plan an energy-efficient transportation/power/irrigation project, taking water from Lake Biwa. Construction began in 1885, and the 1,500,000-yen project took five years to complete. The facility was rebuilt in 1912 to double the amount of electric power generated and to provide drinking water to Kyoto. Properly the Biwa Sosui (sosui = canal), on maps the project also has been called Biwa Canal, Otsu Canal, and Biwa Drainage. It seems best known, however, by the name of the inclined plane, called Keage. The student, Saburo Tanabe, ingeniously provided an efficient means of transportation, an all-weather, high-altitude supply of water for irrigation, a reliable flow of water for Kyoto's industrial and household needs, and a source of energy for conversion to electric power. The project consists of a 5.24-mile lake-level canal, a 0.35-mile inclined plane, a 1.12-mile Kyoto city-level canal, and a 6.61-mile canal alongside the Kamo River—13.32 miles in all from the Lake Biwa shore in Otsu to the Uji River in Kyoto's Fushimi ward.

This project, which put water to use in all its essential ways, was completed in 1890. The Kyoto municipal government owns the canal and incline, and leased the navigation rights to private entrepreneurs.

Lake-level Canal

Mr. Tanabe took advantage of the natural contours to hold level to the extent possible with minor excavation across the Yamashina valley in the middle of the mountains, but had to pierce four intermediate ridges with tunnels bored through rock. Reckoning from the east, the tunnels extend respectively 8,038, 2,067, 328, and 2,854 feet in length, accumulating to 2½ miles underground. The canal is approximately 28½ feet wide at the water line, and its banks are paved with close-fitting stones to thwart erosion, while bridges are of masonry. This canal drops eleven feet over its entire length—on an average 0.04 percent gradient—to assure rapid flow at the volumes wanted. Ever conscious of the environment, the Japanese lined the canal with cherry and maple trees that have become favorite haunts for fireflies. Where the lake-level canal reaches the heights above the city, the project provided a large reservoir and water treatment plant for the public water supply, and a hydroelectric power plant. The canal ends at an inclined plane.

Wooden pointed-prow barges (small by U.S. standards) carried cargo and passengers, floating with the canal's current, passing through the tunnels (like Paw Paw Tunnel on the Chesapeake & Ohio Canal in Maryland), and riding the incline (as on New Jersey's Morris Canal). The westward boat trip with the current consumed one hour. The reverse trip, city-to-lake, required poling of the boats, and took two and one-half hours. Although what could be regarded as a towpath follows the

(Continued on Page 80)

Looking west from the vantage point used for the previous photograph, showing the light-rail transit train crossing to the Midera station. The first mountain barrier is in the background. Regulator gates can be seen beyond the second bridge.

JAPAN'S BIWA CANAL

Looking down the Keage Incline, showing the bridge-type trusses on the cradle car.

(Continued from Page 79)

open canal, it does not continue through the tunnels, and manpower was more abundant than the animal power—as evidenced by wide use of man-powered "rickshaws" and man-powered railways throughout Japan in those days. Dim electric lights lit the tunnels, and the boatmen carried torches.

At the top of the incline is one of Japan's oldest and best-known hotels, the Miyako Hotel—now in the Westin chain—which commands a fine panoramic view of Kyoto, and which uses some of the water to irrigate its gardens. At other places along the canal, water is drawn off to flood rice paddies on terraced hillsides.

Also atop the incline, the project detoured water into a sluice that went by a small ditch and short tunnels to a small reservoir, then gradually dropped down the hillside contours. This Sosui-Bunryu (branch canal) leveled and intercepted the Shira-gawa, the Takano-gawa (a major branch of the Kamo River), and the Kamo River, then turned due south as the Horikawa Canal alongside Kyoto's oldest streetcar line to the Kyoto railroad station. This brought water to Nijo-jo Castle and Nichi-Hogan-ji gardens. It continued south essentially as a drainage ditch to the Kamo River. This stonewall-lined branch canal was not navigable and, in fact, drops over several small cascades.

Keage Incline

From the end of the lake-level canal, an inclined plane employed counterbalanced cradles to haul empty boats up and let full ones down to the city level. The Japanese language contains no precise equivalent of "incline;" keage means "kick-up." However, in more recent times, English has been adopted for such terms, but altered to fit Japanese phonetics—hence inkurain for "incline." Keage survives, as the name of the Kyoto neighborhood near the incline and of the nearby light-rail transit station.

The double-track incline is laid with rails spiked to 8-foot, 3.7-inch gauge. The cradle wheels have double flanges, like pulleys. The incline descends 118 feet in a length of 1,820 feet on a gradient of 6.48 percent; all of its mechanism was underground and was probably water powered in keeping with the efficiency objective. In any event, gravity did most of the work, inasmuch as boats loaded with lumber, coal, farm products, and fish for the city went down and empty ones went up. The large-diameter wheels on which the cable turned are on display at the top of the incline. The incline cars are equipped with a pair of railroad trucks set to the broad gauge. For strength, the designers built the cradle car bodies like truss bridges. The excess water runs down a chute alongside the incline and, at the foot of the incline, joins runoff from the power plant in a boat basin.

City-level Canal

Below the incline, the Lake Biwa water flows into a network of shallow canals. Boats could unload their cargo at docks, or continue to most Kyoto neighborhoods. The water also flows into or is pumped up to level-controlled moats that surround castles, palaces, shrines, temples, royal gardens, and public parks, as well as furnishing water for their pools and ponds.

The top of the Keage Incline, showing a typical canal boat in the cradle.

At approximately 155 feel elevation, the Lake Biwa Canal continues west, north, and west again 1.12 miles to the Kamo River. Part of this section forms the perimeter of Okazaki Park and Zoo, the Art Museum, the Traditional Industry Museum, and the ornamental grounds of the Heian Jingu shrine (itself built at the same time as the canal, taking advantage of the new water supply). The first quarter of this alignment incorporated the Shira River, and the remnant of this river was canalized as a branch to the Kamo River. Just before the juncture with the Kamo River on the main canal, a lock contends with the varying rates of flow of the river. The builders dammed the river here to form a pool. On the other side of the river, the smaller Takase-gawa Canal followed the river's right bank, bringing water to Kyoto's hotel and retail district. City merchants dipped water from sluices with long-handled wooden cups to sprinkle on the pavement as a means to hold dust down in front of their shops; other residents similarly dipped water for their vegetable and flowers gardens. Like the Hori-kawa Canal, this canal was not navigable. South of the railroad station, the Takase-gawa Canal crossed the Kamo River on an aqueduct, and had its water supply recharged from the Lake Biwa Canal, then carried water south to industries in Kyoto's Fushimi ward, famous for distilling fine sake. More recent channeling of the Kamo River for flood control removed the aqueduct. Between the lock and the Kamo River, the Lake Biwa Canal turned southward alongside the river.

Truck of cradle car, showing double-flange (pulley-type) wheels.

Kamo-gawa Section Canal

Proceeding south parallel to the Kamo River along its left bank, the Lake Biwa Canal continued to carry a large volume of moving water. Like the Delaware & Raritan feeder canal bringing water to Trenton, the canal brought an abundant water supply to power or cool industrial equipment in Fushimi ward, where the canal divided to reach as many sites as possible. This working water went through small sluices predominantly to assist in the manufacture of cloth, and into the Kamo River where its flow was vital to the silk industry, which washed bolts in the river and dried them on the stone-paved banks. With its last strength, the water from Lake Biwa powered an electricity generating station before reaching a canalized diversion of the Uji River at Fushimi port. The last, zigzagged section of the canal carrying Lake Biwa's water is named Go-gawa (simply, "man-made river"). Until modern sewer systems were installed, the fast-flowing water also carried sewage away.

Further Transportation Development

A streetcar line climbed the slope alongside the incline to reach the Miyako Hotel about 1895, making the incline redundant for passenger use. After the university student's project had served for a bit more than a decade, its contribution to transportation became less direct. An interurban trolley line surmounted the topographic barrier in 1912, running alongside and occasionally in the middle of the Tokai-do pavement, but using a short tunnel at the summit to gain advantage over the road. The Lake Biwa Canal project supplied the electricity to run the trolley cars. Subsequently, in 1921, the Japanese National Railway relocated its Tokaido main line eastward from Kyoto on a direct route to Otsu through a pair of tunnels—the Higashiyama tunnel and the Second Osakayama Tunnel—with a new station between at Yamashina. Funicular railways have climbed both flanks of 2,783-foot Mount Hiei since 1925 (from Yase-yuen) and 1927 (from Keburu Sakamoto). Boating of goods and people on the canal diminished and eventually vanished.

Still later, the Shinkansen (what we call the "bullet train") penetrated the mountain barrier through its longer, lower Higashiyama and Otowayama tunnels. Then, the Meishin (or Nagoya-Kobe) Expressway joined the other transportation arteries through this narrow corridor. Also, to put the energy of the lake's water to further use, a hydraulic tunnel was bored from the Uji River's source to the river at Uji, where it drives a modern power plant. The river is dammed in the gorge to drive an additional power plant. More recently, in 1974, the Japanese National Railway completed the Kosei-sen railroad (and its own tunnel), skirting the west short of Lake Biwa, and most recently, a lesser highway with tunnel was built in the same vicinity. Through this busy Otsu-Kyoto passage flows much of Japan's commerce.

The Keage Incline, showing the gates for the power plant intake. The railings on the cradle car have evidently been rebuilt.

Looking up the Keage Incline from just above the city-level canal. Tracks at the bottom have been covered with fill for the convenience of hikers.

The canal continues in all its intended uses except carrying cargo and passengers. The municipality maintains the canal property in park-like condition. The incline remains in place, with a wooden boat in each of the two cradles, although the cable has been removed. While it is doubtful that the designers envisioned recreational use initially, it occurs in the simple forms of picnicking on the canal banks and incline grounds, hiking along the berm, swimming, and fishing.

Although the streetcar line was replaced by buses, the interurban trolley line has evolved as a modern light-rail transit line, a branch of which crosses the canal at Midera near the intake from Lake Biwa. The Keihan Tetsudo (Kyoto-Osaka Electric Railway) usurped much of the southward continuation of the canal along the left bank of the Kamo River for a relocation of its interurban line from Osaka off the canal berm into a subway south of Sanjo, much as was done with the bed of the Morris Canal through Newark and the Erie Canal through Rochester. A joint railway company is currently constructing a subway in the canal remnant north of Sanjo to bring the Keifuku light-rail transit lines from the north to the heart of the city. Kyoto's fifth subway line, under construction, will also pass through the intermountain area traversed by the Lake Biwa Canal.

Meanwhile, an unending procession of school children examines the Lake Biwa Canal and Keage Incline for inspiration. The young people observe a 101-year-old successful public works project demonstrating how man and nature can coöperate.

Clever Mr. Tanabe deployed Lake Biwa water in every imaginable way. Governor Kokudo must have been pleased.

(Gordon J. Thompson, 32 Highgate Avenue, Buffalo, NY 14214, is an Urban Transportation Planner and Consultant.)

Those interested in doing further research on the Biwako Canal are referred to the Canal Society of New Jersey's *TOWPATH POST*, Summer 1974 (whole issue), written and illustrated by Bill Trout which describes the canal as it was nearly 20 years ago. The bibliography includes a list of books on the canal in Bill Trout's library, which need to be translated from Japanese! (For further information, contact Bill Trout at 35 Towana Road, Richmond, Virginia 23226.)

A pedestrian bridge over the Lake Biwa Canal at Misaragi in Yamashina, now a precinct of Kyoto.

JOURNEY ON THE DOURO RIVER OF PORTUGAL

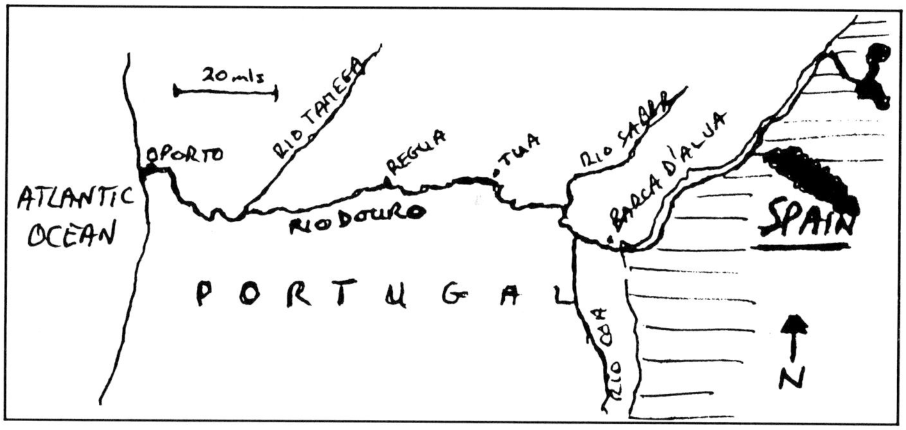

by Roger W. Squires

The great rivers of Europe have been used as navigations since before Roman times. Some, however, had such steep gradients that they could only be used at certain times of the year for commercial navigation. Over the years, one by one, they have been improved by the construction of locks and dams. One of the most recent developments has been the upgrading of the Douro River in Northern Portugal.

The headwaters of the Douro rise at Pico de Urbion in north central Spain and in the border area the river flows through a 75 mile gorge which has always been virtually unnavigable. The head of regular navigation has usually been accepted as Barca d'Alva close to the Spanish border. The river valley became the focus of wine and port production in the 18th century and until the 19th and early 20th centuries the produce of the local vineyards was carried in high-sterned river boats, the baros rabelos, which negotiated the wild waters at great risk. These were guided by oarsmen downstream but generally assisted by horses for the upstream journey. Each craft was capable of taking about 60 tons to the seaport at Oporto. A railway line, which mirrored the river, took away the waterborne trade in the 20th century. Now only an annual regatta at Oporto provides an opportunity to see the barco rabelos in action.

In the early 1960s planning began for a series of five hydro electric dams and locks, the latter being capable of passing 2000 ton barges to transport iron ore from the Moncorvo mines, in the border area, 130 miles downstream to Oporto for transhipment. The taming of the Douro River, which drops 426 ft. in 130 miles, first began in the 1970s with the construction of two locks and dams at Carrapatelo and Regua. The shaft lock at Carrapatelo having a lift of 114 ft. (34.5 m) with the Valeira lock, with a lift of 32.5 m, opening in 1976. Problems with funding delayed the completion of the remaining locks and barrages at Baguste and Pocinho until the mid 1980s. The full navigation was not finally completely operational until 1988.

Our trip started in Oporto where the boat provided the facility of a base whilst we visited the town. Porto, as it is called locally, is of Roman origin. The original settlement being on the south bank in an area now named Vila Nova da Gaia where most of the port (wine) houses are to be found. Three bridges span the river at Oporto whilst the town itself has granite houses which seem to overlap each other on the hillside.

We cast off at lunch time on the second day, after a further mornings wine tasting, to sail upstream for four hours, through the meandering valley with its lush pastures, pines, eucalyptus and almond groves to pass through the Crestuma lock and barrage lake to the town of Entre-on-Rios, where the Tamega River joins the Douro, to moor overnight.

The next day saw an early morning start as we continued upstream with the pines and almonds on the valley sides intermixed with olive groves. The highlight of the trip was the passage through the Carrapatelo Shaft lock that raised our boat 114 ft. in under 20 minutes. We made a brief stoppover to see the hot springs at Caldos de Aregos in Resende, before continuing upstream past the terraced vines, to Peso de Regua where we moored in the late afternoon. Regua proved to be a fascinating town and marks the official boundary of the port wine region.

The next morning we made an excursion to Lamega, south of the river and climbed the seemingly endless steps to the Sanctuary of our Lady of Remedios from which we gained a fantastic panorama of the whole area. At lunch time we reembarked to continue our voyage upstream to Pocinho, passing through the three locks of Baguste, Valeira and Pocinho. This transit took some 7 hours in all.

The fifth day of our trip was spent exploring the region known as Tras-os-Montes (over the mountains), which nestles between the Tamega River and the Spanish border and is the area where most of the port wine grapes are grown. Enroute we saw the old bridge at Mirandela over the river Tua, and the Roman bridge at Chaves which is still in use today. In the late afternoon we reboarded the Pinto do Silva and set sail for Barca d'Alva, the head of the main navigation. Above the town the Douro narrows to pass through a long gorge.

The following morning our boat turned to head downstream to Pinhao. A journey that took about 6 hours. When we arrived we were first taken to the local railway station, where the azulejos depicted various aspects of life in the Douro valley, before visiting a local quinta (farm).

The next day we left early for our final cruise downstream through the remaining locks to Oporto where we moored in the evening light at the old river quay. Sadly, that was the end of our voyage.

We spent the final night on board before returning to the airport the next morning, laden with our samples of port wine, after a journey through the magnificent scenery, picturesque riverine ports, quintas and wineries of the Douro River valley.

(Our trip was booked through Serenissima Travel, 21 Dorset Square, LONDON. NW1 6QG ENGLAND. Phone London 071 723 6556)

ROSCOE VILLAGE CALENDAR

This placid scene (in four colors) of the horse-drawn canal boat at Roscoe Village, Ohio, forms an important part of the beautiful 1992 Calendar just released by the Roscoe Village Foundation. Roscoe Village is a completely restored canal town which today looks as it did in 1830 when it became an important port on the Ohio and Erie Canal.

ROMANCE ON AN ENGLISH CUT[1]

(EDITOR'S NOTE: A. Knacker Nayligh (otherwise known as Pat Saunders of Lamp Post Cottage, Blackheath, Guildford, Surrey, GU4 8RB) has provided this little story to help Americans brush up on their conversational canal English. For those who are a bit rusty, he has included a glossary.)

Veronica stood gongoozling[2] her butty[3] as it winded[4] in the hole.[5]

"Shall I scrumble[6] my cratch[7]?" she pondered, "or smother it with roses[8] and cut down its bulk[9]?" (Her motor[10] was being blacked[11] and re-chalicoed[12] at Charity.[13])

She desparately wanted to impress Harry, the ex-knobstick[14] Number One[15] who was on her tail.[16] She'd shown him her back-end[17] once on the Moira[18] and caught him on the Jackdaw[19] when he'd had to loose her by.[20] He'd even commented on the spring of her looby[21] when she was lying at the top of Johnson's bottom[22] during a stoppage.[23]

So she knew he was quite impressed, but she half suspected he wanted to get his hands on[24] her snubber and snatcher,[25] Willers[26] having better gear[7] than the old Joshers,[28] or even the Gassers.[29] Of course, the Joey's[30] simply had to make do. She'd sometimes had to hold out[31] for a string of them, including 'amptoners,[32] between Factory Three[33] and Claytons.[34]

When times were bad her father, not really a horse-marine,[35] had done a Rodney[36] for a few bob[37] on the thirteen-twelve[38] near the center of Brum,[39] the BCN[40] being the Joey boating stronghold. She'd wheeled[41] for him on the hoss-path.[42]

Her father's family had been boating for many generations but her mother came off the land[43] which was why she was a scholard.[44] She liked to hear his stories of the old days when they'd sometimes loose off[45] and peg to[46] before sparrowshave,[47] or even work fly[48] to beat a stoppage.[49]

As she'd been down the junction[50] working juicers[51] Veronica hardly knew Harry, but had heard tales of his feats, including shafting[52] all night when his Bolinder[53] conked.[54]

In a way she regretted the passing of the horse boats. "I know a motor makes life easier," she reflected, "particularly if its a good lifter[55] and you've got your backend[56] well down, though its harder on the ellum.[57]"

You had to have a good screw,[58] though. Some boatmen would lodge on[59] a sill and hammer the blades[60] to get better fan hold.[61]

"But all the old ways were more pleasant," she thought, "with pairs of animals,[62] the occasional mule or jenny,[63] but most of the time stocky little cobs. Too many hands[64] and the haines[65] would clout the bridge 'ole.[66] The gears[67] with the bobbins[68] and painted spreaders[69] looked good too, better than the leathers[70] and swingle tree[71] of the west countrymen.[72] "Though," she went on, "I wouldn't fancy legging[73] through Blisworth[74] on narrow cut wings,[75] even with a good swimmer.[76] And what about Kit Crewbucket[77]?"

Her reverie was broken by a wheeler[78] on an old Herc.[79] tearing up to set.[80]

"Seen Harry?" she shouted, dropping[81] while he drew.[82]

"Back three bridges lengths,[83]" he yelled back, "and ockering.[84]" The breast posts[85] gaped and she put her back into[86] the beam.

To cut a long story short, when Harry came up with his single[87] they breasted up,[88] and motored off full chat[89] into the sunset. The Sally Ann major[90] spliced them[91] at Suttons[92] over the August Stoppage[93] and they lived happily ever after.

GLOSSARY

1. Canal or artificial section.
2. Gazing idly.
3. Unpowered "narrow boat," 70 ft. by 7, loading about 30 tons.
4. Turned, using the wind to blow it round.
5. Winding hole = turning basin.
6. Apply dark varnish.
7. A pyramid-type structure at the front of the hold.
8. Painted roses and castle were traditional designs.
9. An artificial projection of canvas, largely decorative.
10. Powered "narrow boat." Usually a powered and unpowered boat (butty) worked in pairs.
11. Repainted with bituminous paint.
12. Compound of tar and stable manure used to waterproof wooden hulls.
13. Charity Dock — an English Midlands boatyard.
14. Boats and boaters from the English pottery towns round Stoke on Trent.
15. A boatman who owned his own boat.
16. traveling close behind.
17. Stern, i.e., left him behind.
18. The Ashby de la Zouch Canal.
19. An extremely twisted length of the Grand Union Canal.
20. Allow to overtake.
21. Spring loaded towing hook at the top of the towing mast.
22. Moored above the bottom lock of Johnson's Hillock flight.
23. Canal closure for repairs.
24. Steal.
25. Short and long tow ropes.
26. Craft of Willow Wren Canal Company.
27. Equipment.
28. Craft of Fellows Morton and Clayton Carring Company, founder Mr. Joshua Fellows.
29. Boats carrying tar from coal burning gas works, pre-North Sea gas.
30. Crudely constructed open short-range boats of Birmingham Canal Navigations.
31. steer away from towpath side.
32. 90-foot long boats in the Wolverhampton area which had to pass no locks (locks 70 feet long).
33. A flight of 3 locks at Tipton, 8 miles from Birmingham, 5 miles from Wolverhampton.
34. Thomas Claytons boatyard at Oldbury.
35. A man with a horse who assists unpowered boats.
36. Doing a Rodney = assisting in the passage of a flight of locks.
37. a few shillings.
38. A flight of 25 locks with a short break in the middle. Farmer's Bridge, Birmingham.
39. Birmingham. Also "Birnigum."
40. Birmingham Canal Navigations.
41. Used a bicycle to set locks ahead of the boat.
42. Towpath. Elsewhere "hauling way" or "haling way."
43. Lived in a house.
44. Could read and write.
45. Untie mooring ropes.
46. Connect horse to tow line.
47. Daybreak. The term "sparrowshave" is used here to spare American sensibilities. The term in actual usage is "sparrow fart," a metaphor based upon ornithological indigestion.
48. Go non-stop.
49. Maintenance closure.
50. The south end of the Grand Union Canal.
51. Boats carrying barrels of concentrated lime juice.
52. Pushing with a pole or shaft.
53. Early simple diesel engine.
54. Ceased to work.
55. Works well under load.
56. Stern.
57. Rudder.
58. Propeller.
59. Perch the stern of the craft on the sill of a half-empty lock to improvise a dry-dock.
60. Hit the individual propeller blades to increase the pitch: very pragmatic.
61. Grip on the water.
62. Donkeys.
63. Mule: A hybrid with donkey daddy and a horse as mummy. Jenny: A hybrid with role reversal (or vice-versa - who cares? Not even another mule (or Jenny)).
64. Too tall: a hand = 4", at the shoulder.
65. Haines: Part of harness, also known as hames, aynes, whums.
66. Clout: hit (not "clart," which is horse manure in the NW). Bridge 'ole: A low narrow bridge over the canal.
67. Harness.
68. Painted wooden rollers with rope traces running through.
69. Wooden bar at stern of horse gear to attach towropes via a spliced rope.
70. Leather traces.
71. Similar to 68 but with a hook to slip an eye splice on the towrope over.
72. Barges working west from London (Not in Devon and Cornwall).
73. Propelling boat through tunnels with no towpath by lying on one's back and "walking" along the walls.
74. Canal tunnel 1-1/2 miles long.
75. Narrow cut wings placed across bows of boats for leggers to lie on.
76. Moves easily through the water.
77. Fabled ghost who haunted a tunnel.
78. See 41.
79. "Hercules" brand of bike.
80. Get lock ready.
81. Lowering "paddles" to close water flow through top gate.
82. Open paddles to drain lock.
83. A canal boatman's unit of distance. Three bridges lengths may mean 100 yards or two miles. Similar to American measurement in "blocks." Do I walk three blocks or get a cab?
84. Hurrying.
85. The mitered vertical beams where two lock gates meet at a point.
86. Pushed hard with her shoulder blades and braced legs.
87. Motor boat with no butty.
88. Lashed side by side.
89. Very fast.
90. Salvation Army mission worker to boat population.
91. Married them.
92. Sutton stop lock. This has a few inches rise at the junction of the Oxford and Conventry canals.
93. See 49.

About the author: A. Knacker Nayligh was named after Enoch and Eli, two mythical Birmingham and Blackcountry (BCN) characters, the Abbot and Costello of homely wit, e.g.:

Enoch (who has bought a mongrel dog): "Cop this, Eli."

He throws a stick across the cut. The dog walks across the surface of the water and retrieves it.

Eli: "You paid a quid for 'im? You bin don."

Enoch: "Why'm I bin don?"

Eli: "Booger can't swim!"

This article written for us by Pat Saunders was one of his last, for shortly after his return to England he died of a stroke at age 66. Pat was one of the pioneers of the canal restoration movement in Britain. We'll all miss him.

MONTECH WATER SLOPE REVISITED

By Roger Squires

From the earliest times man has sought ways of moving boats uphill. The ancient Egyptians and later the Chinese used mud covered inclines as a means of moving boats between two levels. However, it was finally the French who devised a method whereby a boat could actually float in a channel whilst going uphill.

The idea of the true 'water slope' was developed by two French Waterways Engineers at the Chatou Laboratory in the mid-1960s. They built a one tenth scale model at Venissieux near Lyons in 1967 and after extensive trials were given the go ahead to build a full size prototype at Montech on the Canal Lateral a la Garonne in 1969. (Near Toulouse in southwestern France.) This was part of an upgrading scheme for the canal so that it could pass 38.5 m barges. The site was chosen because it offered an economical way of replacing 5 closely spaced shorter locks. The construction was undertaken by SPIE Batignolles under the supervision of the French Ministry of Equipment, Navigation Service. The works were completed at the end of 1972 and after extensive trials the Montech "Water Slope" was opened to navigation in 1974.

A tour boat being shoved up the Water Slope.

The concept is very simple, but relies on modern technology in two respects. One is the roller seals that enable a large ram to fit snugly into the flume channel, yet move along it without leaking water. The other is the electrical synchronisation between the two drive motors situated either side of the flume channel that are used to push the ram up or down the water slope. At Montech the chosen method of propulsion was a matched pair of 1000 horse power diesel engines each mounted separately on four rubber tyred traction axles. These axles straddle raised concrete guide rails on concrete road beds either side of the flume trough. The boat floats on a wedge of water in front of the ram which is itself cantilevered between the two traction motors. Both the motors are also interlinked with a beam structure through which the synchronisation cableducts pass.

The system of operating is quite simple. Apart from the moving ram only one stop gate is required at the upper level to retain the water in the higher pound. On the basis that the craft wishes to transit up the slope, it enters the lower section of the flume channel on the level. Once it has passed under the raised traction ram, the ram is lowered and the boat secured fast to bollards on either side of the ram platform. Pressure rollers on both the ram sides and the base make a watertight seal. On a signal from the Engineer the boat, now secured to the ram, goes into forward gear and the traction engines start to move the ram with its wedge of water up hill. The boat floats on this wedge with its forward drive keeping its rudder clear of the ram. When the upper level is reached the ram automatically halts. The stop gate in front of the boat is lowered. The boat casts off and proceeds along the upper level of the canal. At Montech this whole process is completed in seven minutes for the vertical rise of 13.3 metres. For a boat descending much the same drill applies. In this instance the bow of the boat is secured by two side bow ropes to the padded front of the ram. The stop gate is raised behind the boat to seal off the upper level. Once this is done the ram, with boat in forward drive to keep it snug against the ram fenders, proceeds down the slope floating on its wedge of water. At the bottom of the slope, the boat is untied. The ram is raised and the boat sails off along the lower level. The whole process is controlled by a single engineer who rides on one of the traction units in a railway loco style cab.

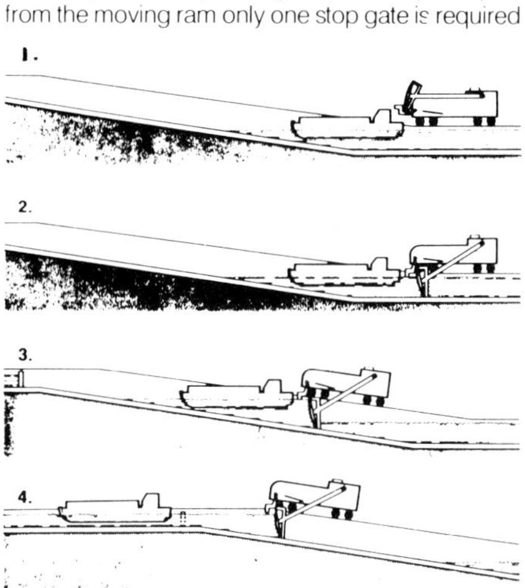

Uphill sequence at Montech "waterslope": (1) Boat on lower level passes beneath raised push-plate and enters approach basin. (2) Push-plate is lowered behind boat. Rolls covered with Du Pont Neoprene rubber seal three edges of plate against leakage. (3) Wedge of water thus created is pushed forward by parallel locomotives on each bank. (4) At top of slope, retaining gate topples to allow rising wedge of water to merge with upper water level. Boat moves forward under its own power and push-plate returns to its initial, raised position.

For the technically minded the flume channel of the water slope at Montech is six metres wide and four hundred and forty three metres long and rises on a 3 degree slope. The traction drives move at 4.5 kms per hour. Because of the nature of the ram/boat connection, the slope at Montech can only pass full width peniches. This is because the ram to boat links are limited to two side ropes. A narrower boat could twist in the channel and this could cause distortion to the ram and thereby create a water leak. I travelled the slope in a peniche twice in mid 1993 and the water seals were very sound. I saw no deterioration from when I previously visited the slope in 1981. Such is the reliability of the modern materials used is the ram. However, commercial traffic has now ceased on this section of canal and this itself could place in question in the longer term future of the Water Slope.

The early success of the Montech water slope prompted the French to build a more sophisticated electrically powered version at Beziers on the Canal du Midi to bypass the staircase of six locks of Fonserannes as part of a scheme to open up the canal to 38.5 metre craft. Work was started here in 1981 and completed in 1983 at a cost of some £3 million. At this site the moving ram system was provided with long side mooring bars so that it could pass pleasure craft as well as the 38.5 metre barges. The unit was designed with a capacity of transporting up to 14 peniches and 200 pleasure craft in a 13 hour day. Sadly that capacity was never achieved due to the decline in commercial traffic. Various technical faults, including a 'runaway' , also have meant that the slope seems to be more often under repair than in operation. This was the case when I visited it in June 1993. However, when it does work it gives pleasure and commercial craft alike a very speedy transit over the 15.5 metres lift. This is most valuable at what often becomes a bottleneck on the canal.

Even if the modern electronics at Fonserannes do create problems, the Water Slope is a very novel way of moving boats uphill. The speed of the operation of the Montech Slope has much to commend it even today. It is certainly well worth a visit if only because of its novelty value. A ride on it is in my view a 'canallers' dream.

(Passenger trips at Montech can be organized through the owners of 'Baladine', 2788, Chemin Du Quart, VERLHAGUET, 82000 MONTAUBAN, FRANCE. Phone: 63 64 76 64.

The cars control the push-plate and water wedge.

INDEX

THE AMERICAN CANAL SOCIETY INCORPORATED

The American Canal Society was incorporated in Maryland in 1972 as a "non-profit scientific and educational organization". Its objectives are: to encourage the preservation, restoration, interpretation and usage of the navigation canals of the United States, past and present; to cooperate with individual canal societies for action on threatened canals, or in the absence of local canal societies to act as a focal point for action; and to provide for the exchange of general canal information. It is not our intention to interfere with the autonomy of individual canal associations or societies in the conduct of their own affairs, but simply to fill a gap in providing a clearing house and distribution medium for canal information and activities.

In addition to publishing our quarterly newsletter "AMERICAN CANALS", we are also in the process of publishing a series of AMERICAN CANAL GUIDES, which will ultimately cover the entire country. Five have been produced to date: (1) The West Coast (2) The South East (3) Lower Mississippi and South Central (4) West Virginia, Kentucky and the Ohio River (5) Virginia, Delaware and Maryland.

Present Officers of the Society are as follows:

Full information on membership requirements, dues, available literature, etc., may be obtained by writing:

Charles W. Derr, Secretary-Treasurer
117 Main Street, Freemansburg, PA 18017

OTHER PUBLICATIONS OF THE AMERICAN CANAL AND TRANSPORTATION CENTER

THE CANALS OF NEW YORK STATE – A publication of the American Canal Society (1991). The canals of the Empire State, past and present, are covered in whole or in part. 8-1/2" x 11" paperback, 48 pages, 85 illustrations.

THE AMAZING PENNSYLVANIA CANALS – 160th Anniversary Edition – By William H. Shank (1991). A much expanded variation of many previous printings. 125 illustrations, and tables of locks and mileages on most of the principal canals in the State, never previously gathered together in one volume. Four-color cover; two-color interior; 128 pages; a definitive work.

TOWPATH GUIDE TO THE CHESAPEAKE AND OHIO CANAL – By Thomas F. Hahn, (1990). A fully illustrated, historical commentary and mile-by-mile directory for the entire 184-mile length of the C. & O. Canal Towpath from Washington, D.C. to Cumberland, Maryland. Excellent maps included. 226 pages.

HISTORIC BRIDGES OF PENNSYLVANIA – By William H. Shank, (1986 Edition). Traces the development of the bridge-building arts from the time of the first covered bridge in America, to modern bridges of the 20th Century. Biographies of such famous bridge builders as John Roebling, Theodore Burr, Charles Ellet and Ralph Modjeski included. Profusely illustrated.

PENNSYLVANIA TRANSPORTATION HISTORY – A Supplement – By William H. Shank (1990). In this book, Mr. Shank has discussed media and devices not covered in his other books. Early river craft, rope ferries, steam boats, inclined planes, gravity railroads, early steam locomotives, horse cars, cable cars, trolley cars, elevated rail and subway systems, and air-travel devices are included. An 8-1/2" x 11" book with 72 pages and approximately 100 old photos, drawings and tables. Four-color cover.

GREAT FLOODS OF PENNSYLVANIA–A TWO-HUNDRED YEAR HISTORY – W. H. Shank, (Sixth Printing, 1988). Data, photos and non-technical text on all major floods in the Keystone State since records have been kept. Full chapters on the Johnstown Flood and the disastrous floods of 1936, 1955 and 1972 are included. A definitive work.

INDIAN TRAILS TO SUPERHIGHWAYS – By William H. Shank, (1988 Edition). History of the development of Pennsylvania's historic roads and the many interesting vehicles used on them. Descriptions of Braddock's Road, Forbes' Road, National Highway, Lancaster Turnpike, Plank Roads, Corduroy Roads, William Penn Highway, Lincoln Highway, Pennsylvania Turnpike and Keystone Shortway. Recent PennDOT plans for completion of Pennsylvania's Interstate Highway System.

THE BEST FROM AMERICAN CANALS – Number 1 (1972-1979). A publication of the American Canal Society, (fourth printing 1988) featuring articles written by canal-buff experts on historic and currently operating canals in the U.S.A., Canada and Europe. An 8-1/2 x 11 paperback, with 88 pages, the book contains 150 illustrations and a world-wide index.

THE BEST FROM AMERICAN CANALS – Number 2 (1980-1983). A publication of the American Canal Society (1990) reprinting articles on historic and currently operating canals in USA and overseas. Featured are the Panama Canal, the Trent-Severn Waterway, and canals in Europe, Thailand and China. 140 illustrations and maps included, and a worldwide index. An 8-1/2 x 11 paperback with 88 pages.

THE BEST FROM AMERICAN CANALS – Number 3 (1983-1986). A publication of the American Canal Society (1986), featuring the Canals of New England, metropolitan New York area, Erie Canal, the I. & M. Heritage Corridor, the Tenn-Tom, St. Lawrence Seaway, and Canals of Sweden, Germany and India. 145 illustrations and maps, and a world-wide index. An 8-1/2 x 11 paperback with 88 pages.

THE BEST FROM AMERICAN CANALS – Number 4 (1986-1988). A publication of the American Canal Society (1989), featuring maps of the principal canals of the USA, Erie Canal, Inclined Planes, Morris Canal, S. & T. Canal, Shinnecock Canal, Champlain Canal, Navigation on the Mississippi, Kanawha, Kentucky , Fox, and Muskingum Rivers. 8-1/2 x 11 paperback, 88 pages.

THE BEST FROM AMERICAN CANALS – Number 5 (1989-1991). A publication of the American Canal Society (1991) including a complete history of the Panama Canal, up-date on the Rhine-Main-Danube Canal, description of several Chinese canals, and navigation tips on the Black River, the Hennepin, the Alabama and the Florida waterways. 8-1/2" x 11" paperback, 88 pages.

HISTORY OF THE YORK-PULLMAN AUTOMOBILE, 1903-1917 – By William H. Shank, (1970). History of the "Six-Wheeler" Pullman, and its successors, which almost made York, Pa. the automotive capital of the United States. History of the early automotive industry in Eastern Pennsylvania also included. Profusely illustrated.

TOWPATHS TO TUGBOATS – A History of American Canal Engineering. By Shank, Mayo, Hahn and Hobbs (Fourth Printing, 1992). The works of such famous Canal Engineers as Benjamin Wright, Canvass White, Charles Ellet, William Hamilton Merrit, George Washington Goethals are detailed – with the canals they built. The Erie, the Welland, the "Soo," the Panama, the St. Lawrence Seaway and the Tenn-Tom are among the many waterways described in detail. A 72-page, 8-1/2" x 11'" book, the publication contains more than 130 drawings, maps and photographs in USA, Canada and overseas.

WHEN HORSES PULLED BOATS – By Alvin F. Harlow. Second reprinting, 1987, of Harlow's little-known canal book, written for school students in 1936. An excellent exposition of the historic canal era in the USA. Introduction by William H. Shank. Canal bibliography included. Well illustrated with sketches by Orson Lowell and Philip Hoffmann.

VANDERBILT'S FOLLY–A HISTORY OF THE PENNSYLVANIA TURNPIKE – W. H. Shank, (Tenth Printing, 1993). The railroad war of 1880-85 which created the tunnels and roadbed for the present turnpike. History of the Turnpike, 1940-1993, included.

THE COLUMBIA-PHILADELPHIA RAILROAD AND ITS SUCCESSOR – William Hasell Wilson, 1896. This book is an on-the-spot account of the building of one of the oldest railroads in America by its chief engineer, later resident engineer for the Pennsylvania Railroad, who purchased it from the Pennsylvania Canal Commissioners. This 1992 reprint is fully illustrated with 45 old photos, maps and drawings. (Second printing).

THE CANALLER'S SONG BOOK – By William Hullfish, Music Instructor, Writer, Singer (Third Edition, 1993). This 88 page, 8-1/2 x 11 book contains forty historic canal ballads (words and music) collected from old records in the northeastern USA over a ten year research period. Illustrated with old canal drawings. Said to be the most comprehensive collection of canal songs ever published.

PICTURE-JOURNEY ALONG THE PENNSYLVANIA MAIN LINE CANAL, 1826-1857, By Philip J. Hoffmann, P.E. (1993). Edited by William H. Shank, P.E. Full-color drawings of entire state-owned route, Philadelphia to Pittsburgh. 8-1/2" x 11" paperback. 80 pages. Full Hoffmann biography included.

Inquiries may be directed to the American Canal and Transportation Center, 809 Rathton Road, York, PA 17403. Price list and discount schedule available.